P9-APG-323

Technicolor

Technicolor

Race, Technology, and Everyday Life

EDITED BY

Alondra Nelson and Thuy Linh N. Tu
with Alicia Headlam Hines

New York University Press
NEW YORK AND LONDON

NEW YORK UNIVERSITY PRESS
New York and London

Library of Congress Cataloging-in-Publication Data
Technicolor : race, technology, and everyday life / edited by Alondra Nelson and Thuy Linh N. Tu with Alicia Headlam Hines.
p. cm.
Includes bibliographical references and index.
ISBN 0-8147-3603-3 (cloth : alk. paper)
1. Technological innovations—Social aspects—United States. 2. Minorities in technology—United States. I. Nelson, Alondra. II. Tu, Thuy Linh N. III. Hines, Alicia Headlam.

T173.8 .T34 2001
303.48'3—dc21 2001018049

New York University Press books are printed on acid-free paper, and their binding materials are chosen for strength and durability.

Manufactured in the United States of America

10 9 8 7 6 5 4 3 2 1

Contents

Acknowledgments

There was a universal groan in the seminar room the night we learned that we would have to participate in a group research project as part of our degree requirements. Too many experiences with group projects gone bad made us all reluctant, to say the least. But with Debra Wexler Rush we took a leap of faith, and over the course of the year we learned that working together could be as rewarding as it was trying. It was Andrew Ross in that American Studies core seminar who taught us the importance of collaborative work, and it was also he who suggested that we turn our project into what would become *Technicolor*. We thank him for ceaseless encouragement, intellectual direction, and good humor.

From conception to completion, this book has been an entirely collective project. We are grateful to all the contributors for supporting us in our vision while constantly challenging us with their ideas. Tricia Rose's contributions extend well beyond the manuscript. A teacher in the true sense of the word, she imparted to us her dedication to rigorous scholarship and commitment to social change.

In the many offices at New York University that we called home, we could always count on a stimulating chat or the use of a conference room. We thank the students, staff, and faculty at the American Studies, Asian/Pacific/American Studies, and Africana Studies programs, especially Manthia Diawara, Lisa Duggan, Anthony Escobar, Alyssa Hepburn, Kelli King, Risa Morimoto, Nikhil Pal Singh, and Jack Tchen, for both the intellectual and physical space needed to complete this manuscript. Eric Zinner, Daisy Hernandez, and Cecilia Feilla at New York University Press were patient and helpful throughout the editorial process. Thanks also to Mad Mike Banks and Submerge Records for allowing us to use their artwork.

Ideas always take shape through dialogue; this book is indebted to conversations with many friends and colleagues, including Laura and Andre Canty-Swapp, Beth Coleman, Tim Haslett, Logan Hill, Selwyn Hinds, Mark Hines, Adria Imada, Jennie C. Jones, Michelle Kim, Anthony Ng, Mimi Nguyen, Shaifali Puri, Debra Wexler Rush, Nichole T. Rustin, Laura Sullivan, and Ben Williams.

The largest debt, one that we could not possibly repay, is due to our families, Kadija Ferryman, Andrea Hall, Anthony Nelson, Robert S. Nelson, Jr., Ai Linh Tu, Duy Linh Tu, and especially our parents, Mavis E. Headlam, Elaine and Elbert Hines, Delores Y. Nelson, Robert S. Nelson, Sr., Thi Thi Nguyen Tu, and Xuan Van Tu, for their love, support, and many sacrifices.

Introduction
Hidden Circuits

Alicia Headlam Hines, Alondra Nelson, and Thuy Linh N. Tu

The pronouncements of a color-blind future that characterized the early days and utopian impulses of the digital revolution are giving way to more pragmatic discussions of race and technology. The "digital divide" has become popular shorthand for the myriad social and cultural factors that shape access to technological resources. The last few years have seen the release of a report from the Commerce Department on the status of race, class, and technology in the United States, the airing of a PBS film documenting how race and gender have come to shape the ways computer technology gets used and by whom, and the establishment of the digital divide as a regular topic of political debate.

Among the strategies that have been developed to bridge this gap in technological access are the creation of community technology centers (CTCs) that provide low-cost access to technology in comfortable and convenient environments, and the free (though usually time-restricted) access to information resources through the Internet and the World Wide Web at public libraries. But how will we know when the digital divide has been mended? When every poor and working-class family has a computer at home? And how much technological access amounts to equality?

Solutions to the digital divide often fail to address problems that we can't solve by simply placing a computer in every home or classroom, problems that include social barriers, subtle hurdles that look more like glass ceilings than impassable divides, and the dearth of content that reflects the experiences of all potential users. Moreover, the digital divide has become a self-fulfilling prophecy, confirming that people of color can't keep pace in a high-tech world that threatens to outstrip them. The divide—unbridgeable, unequivocal—unwittingly confirms that poor and working-class people of color have a technophobia that's hard to shake.

But there are other theories about race and technology, some of which run counter to the current prognosis, and others that confirm it. *Technicolor* reveals the many hidden circuits that link race and technology. Not content to wring our hands about the digital divide, we sought out the many interfaces where technology and race intersect.

Theorizing Technology

The cultural and social significance of new information technologies has been a constant topic of debate for scholars and cultural critics since the 1980s.[1] In just a few decades, these debates have produced an extensive body of literature and social commentary; so much, in fact, that it is often difficult to avoid inflated notions of how technology has impacted our lives, for better or for worse. But there have always been critics who move beyond the two poles of high-tech extremism—technophilia and technophobia—to more realistically assess lived experience in an increasingly technological world, including such important issues as access, economic and social inequalities, cultural identity and subjectivity, gender, and class.

Scholars of work and labor have pointed out that current technological developments have done little to address the recurring economic disparities produced by a capitalist economy.[2] Some worried that the American labor force would see a "jobless future" in which meaningful work is phased out by high-tech restructuring.[3] Many feared that technology would bring about heightened levels of exploitation on a global scale, most notably in the form of downgraded service sector work in the centers of capital and export processing in their peripheries. Manuel Castells theorized a "network society," in which the expansion of an information economy in powerful countries depends on the work of those who labor on the geographic periphery but whose efforts are centrally important.[4]

But as feminist studies of technology have revealed, technological inequalities are not simply economic issues. Narrators of the information revolution have regaled us with tales of hackers and geeks, and in the process have constructed technology as a site of white male superiority. Cyberfeminists have worked hard to reveal the hidden relationships between women and technology.[5] In her important work *Zeros and Ones*, Sadie Plant rewrites the computer's originary history and argues that computational logics are indebted to women's weaving in the nineteenth century. She also creates a new avatar of technology, Ada Lovelace.[6] But even among these thoughtful witnesses who analyze how technological artifacts circulate in the social world, there is a strange silence around how the experiences of people of color might recast technocultural theory.[7]

Most often when attention is turned to the implications of race for theorizing technology, people of color are cast as victims. We witness this most commonly in discussions about the digital divide, which characterize people of color—predominantly blacks and Latinos—as victims of either economic and educational constraints, cultural priorities, or their own fiscal irresponsibility. And historically people of color have been casualties of technologically enabled systems of oppression, from colonial expansion, to the racial sciences of craniology and phrenology, to surveillance and information gathering.[8]

These narratives of victimization demonstrate that technologies and racism have a long and complex history. But unfortunately these accounts sometimes become rationalizations for why people of color fail to have "productive" relationships with technology, and justifications for the still uneven distribution of technological resources and knowledge. After all, if people of color are seen only as victims, then there is very little reason to entrust them with the tools of the future. But to see technology as only a problem for people of color is to present but one aspect of a multifaceted story. *Technicolor* moves beyond the binary logic that insists that race and technology are always at odds with each other.

New Narratives

Technologies have become ubiquitous in our everyday lives, leaving very few sectors untouched. They have become integral to such commonplace activities as shopping and banking, and they have irrevocably transformed interpersonal communication. In their banality these artifacts have been integrated into our everyday lives.

The theorists of everyday life, like Henri Lefebvre and Michel de Certeau, while not directly addressing issues around technology, have provided an important model for understanding our daily experiences.[9] Everyday life theorists broke with the traditional Marxist focus on work and the laborer by asserting that human experience must be considered in its totality; they insisted that work and leisure operate dialectically with each other. The study of leisure allows one to see people's own critiques of their working condition. In a world that elevates productive labor above all other realms of activities—family life, personal life, leisure—people's ability to "distract," "entertain," or "compensate" for the difficulties of their work lives is seen as a critical response.[10] For Lefebvre, leisure is in fact a social need:

> [M]odern industrial civilization creates both a *general need for leisure* and differentiated *concrete needs* within that general framework. . . . In response to such new needs, our civilization creates techniques which nevertheless have an "extra-technical" meaning and character. It produces "leisure machines" (radio, television, etc.). It creates new types of play which transforms the old ones, sometimes conflicting . . . sometimes overlapping.[11]

In this way, leisure becomes its own productive force; its products and practices offer the means to move beyond a life of alienation. The everyday life approach provides an important theoretical model for understanding both the ubiquity of technologies in our daily lives and the multifaceted ways we use them. As Lefebvre himself notes, technologies are imbued with an "extra-technical" meaning, and it is precisely the search for that meaning that we are concerned with here.

For us, understanding this extra-technical meaning requires understanding how race operates in everyday life. Race, in interplay with gender, class, and sexuality, is a crucial category by which people define themselves, form their communities, and reflect on their cultural histories. Racial communities bring these reflections to bear on their use of technologies in work and leisure, and their approaches reveal more than just an antagonistic relationship to technoculture.

Technicolor presents a full spectrum of stories about how people of color produce, transform, appropriate, and consume technologies in their everyday lives. In order to locate these stories, we found it necessary to use a broader understanding of technology, and to include not only those thought to create revolutions (e.g., information technologies), but also those with which people come in contact in their daily lives. For when we limit discussions about technology simply to computer hardware and software, we see only a "digital divide" that leaves people of color behind. Casting our nets farther and wider allows us to more fully realize the different levels of technical knowledge and innovation that individuals and communities bring to their work, play, and creative expression.

It is incomplete, however, to look only at how technologies are used; here we also ask to what ends. As these essays reveal, contests around technology are always linked to larger struggles for economic mobility, political maneuvering, and community building. While we refuse essentialisms (i.e., the idea that people of color inherently use technologies differently than the majority), we do recognize that individuals and communities employ technologies for very specific goals, linked often to their histories and social locations. The essays here show that for disadvantaged communities, technologies have been used to address historical exclusions and continuing inequalities—sometimes to offer more democratic alternatives, other times to manufacture profit, most often simply fill a need. This is of course not to suggest that these are all progressive undertakings. Contradictions abound here; often lines blur between a self-interested search for pleasure and profit and a movement for more equitable distribution of technological resources and knowledge.

And certainly we do not want to suggest that people of color all share the same relationship, historically or structurally, to technology. Techno-savvy Asian whiz kids, for example, have always had a place in the high-tech hierarchy. While quick scans of any registry of media moguls reveal very few nonwhite male faces, the Asian presence in programming, management, and data processing is undeniable. Hardly at odds with technology, as implicitly is the case with blacks and Latinos, they seem the heirs apparent to the technological revolution. Yet to acknowledge the differences is not to understate the commonalities; people of color share histories of racial oppression and disenfranchisement that can be exacerbated by new technologies.

Nowhere is this more apparent than in the technological workplace, where superexploited labor is used to create the tools that promise freedom. This collection looks at three crucial sites of high-tech labor—production, programming, and content/design—to make known the many levels of work with which people of color are involved. Examining these sites also reveals the additional vulnerabilities to which they, particularly women of color, are subject.

In her essay, Karen J. Hossfeld makes visible the hidden labor of immigrant Asian and Latina women in technological production, calling attention to the high costs of progress. Women in Silicon Valley have long been at the forefront of production, while having very little access to technologies as spaces of creative expression and self-representation. But at the same time that these women are the obvious casualties of our technological revolution, they also employ various workplace strategies in order to resist their daily exploitation. South Asian programmers, Amitava Kumar argues, share a similarly troubled relationship to their work. These immigrants have parlayed their technical skills into U.S. visas, but find that life here is not what they expected. At the same time that they enjoy the benefits of programming work, including relatively high pay, they also frequently find themselves mired in a morass of immigration policy, intraracial class politics, and labor exploitation.

While these workers are the hidden circuits of technical labor, the most vaunted figure in new media is the digital entrepreneur. But McLean Mashingaidze Greaves, who counts himself among this group and whose company Virtual Melanin has been creating urban content for many years, shows that even technological work at this level is not free of barriers and stereotypes. In their interview, Andrew Ross and McLean Greaves explore the limits and possibilities of entrepreneurship and the creation of communities of color on the Web. They discuss the tensions between creating ethnic content that reflects the needs and desires of a particular community and the profit-making imperative of the entrepreneur. These tensions reveal that interventions by people of color are rarely purely oppositional. Yet what Greaves's work does is to trouble the long-standing myth that abstractness in cyberspace is both possible and desirable (though, admittedly, that particular fantasy has recently given way to a belief in the importance of differentiation). But perhaps more important, sites like his give commonly underserved communities a reason to log on and to create an online presence bound by shared interests and racial identities.

Mimi Nguyen shares Greaves's ambivalence (if not his profit-making goals) about the promise of abstract citizenship offered by the Internet. Inspired by the same desire to assemble an online community she could not readily find, Nguyen started an Asian American feminist Web site. But she soon discovered that when she made her identity visible by narrating her physical body into the supposedly disembodied zone of cyberspace, she was accused of being impolite and agitational. Here she discusses the transition of her 'zine *Slant* from its low-tech Xerox version to its high-tech Web version and reveals how race and gender implicitly function in this public arena.

While "virtual communities" have long been the Web's most obvious selling point, less often acclaimed are the "real" communities that have formed around struggles for access to and enjoyment of technologies. Here Logan Hill looks at both the universal access movement, taking place on Capitol Hill and in the pages of national news media, and the local efforts of community technology centers. It is often assumed that we can easily remedy the problem of unequal access by placing computers in low-tech communities, but Hill shows that barriers to access operate on many levels and therefore solutions must take multiple approaches. He sees great promise in a variety of efforts: allocating federal funding for computers in schools and libraries; empowering individuals through technological training, mentorship, and job opportunities; and encouraging online participation by the creation of racial/ethnic Web sites.

Complementing Hill's observations, Guillermo Gómez-Peña and Beth Coleman note that exclusions from technological resources and skills often happen tacitly. Gómez-Peña stresses that, along with a dearth of hardware, one of the main barriers to Chicano participation on the Web is language. For the many who do not read and write English, the Web is incredibly difficult to navigate. Gómez-Peña emphasizes the importance of discursive interventions into cyberculture: "What we want is to 'politicize' the debate; to 'brownify' virtual space; to 'spanglishize the Net'; to 'infect' the lingua franca." Only then can we combat "tecnofobia" and the commonly shared assumption that Latinos lack the intellectual capabilities to deal with technology.

In her discussion with Tricia Rose, Beth Coleman talks about a similar unspoken—and often unrecognized—process of exclusion. It is often the case that technical skills are gained through casual use of technologies and through networks of informal "apprenticeships," where friends demonstrate to each other their techniques. This is particularly true with music technologies, in which the types of knowledge needed to create innovative sounds are rarely taught in formal education. For women, however, it can be extremely difficult to gain entrance into the social spaces where this type of technological knowledge is shared, like the music studio, because, for a variety of reasons, it is implicitly a male space. This form of gendered exclusion has had profound effects on women's participation in music production and in DJ-ing.

While access to technology remains one of the most pressing obstacles for people of color, they have often overcome this challenge by making do with what they have. By "refunctioning" old/obsolete technologies or inventing new uses for common ones, communities in many places have fashioned technologies to fit their needs and priorities.[12] In the process, they have become innovators, creating new aesthetic forms (art, music), new avenues for political action, and new ways to articulate their identities.

Ben Chappell shows that the practice of lowrider customization is in fact a highly technical process, requiring a great amount of skill. By refunctioning the car to provide maximum visibility, Chicanos in Texas and California are asserting their own ideas about the preferred function of the automobile—to serve as an object of beauty and display. The adornment of the car to reflect their urban, working-class Mexican American identity takes precedence over alterations for speed and efficiency. Not only do lowrider modifications mess with the cars themselves, they also trouble ideas about the "appropriate" uses of the automobile.

The documentary filmmaker Vivek Bald urges people of color to ignore ideas about what is appropriate and to instead "appropriate technology" for their own purposes. In his two most recent projects, *Taxi-vala/Auto-biography* and *Mutiny: Asians Storm British Music,* he demonstrates how working-class South Asians have made do and made new their everyday technologies. In the former, the CB radio, a throwback trucker gadget made famous in the 1970s by *BJ and the*

Bear, has been retooled as an instrument that facilitates community formation and political action among South Asian cab drivers in New York City.[13] In the latter, simple music technologies—at least by today's music production standards—become tools by which British Asian youth gain technical and artistic skills. At youth centers in East London, funded in part by successful Asian bands like Asian Dub Foundation, Asian kids are taught how to use sampling and DJ-ing equipment and encouraged to utilize music to express their political voice. The brand of cultural politics stressed at these centers is crucial for youth who often have very little access to formal channels of political protest.

Chinese immigrants in the metropolitan New York area use another type of music technology—the karaoke machine—for other purposes: as a tool of pleasure and social activity and as a means by which to communicate across class and regional differences. Karaoke participants draw from and reinterpret "traditional" Chinese music and American "classics" to tell stories about their specifically Chinese American experiences. For Chinese in New Jersey, Queens, and Chinatown, the karaoke "scenes" allow them to blend memories of their past with their contemporary lives, making the formation of a diasporic identity possible.

In deindustrializing Detroit, African Americans articulated new black diasporic subjectivities expressed through technological innovations. They fashioned techno, an electronic musical form, by assembling a variety of sources: African American musical history, the economic and cultural history of Detroit, and a global music import culture to create a unique sound. In their music, techno-pioneers like Juan Atkins, Derrick May, and Kevin Saunderson married myths and technical skills to create what Ben Williams identifies as "black secret technology."

Combined, these essays provide a much-needed first step toward a fuller understanding of race and technoculture. Of course they are not meant to be definitive or comprehensive; rather, they are suggestive of some of the more relevant issues in this emerging field. Here we have tried to point out some of the most pressing problems, including workplace inequalities and racial divides in technological access, as well as some of the more promising interventions. By looking at everyday life and a wider range of technologies, we are able to affirm our roles as producers and innovators, as members of various technological communities, and as participants in the creation of a wired world.

NOTES

1. While the large volume of literature on this topic prevents us from making an exhaustive list, some texts that have shaped debates about technology in the social world include Stanley Aronowitz, ed., *Technoscience and Cyberculture* (New York: Routledge, 1996); Michael Benedikt, ed., *Cyberspace: First Steps* (Cambridge: MIT Press, 1991); Gretchen Bender and Timothy Druckery, eds., *Culture on the Brink: Ideologies of Technology* (San Francisco: Bay Press, 1994); Wiebe E. Bijker, *Of Bicycles, Bakelites, and Bulbs: Toward a Theory of Sociotechnical Change* (Cambridge: MIT Press, 1995); James Brook and Iain Boale, eds., *Resisting the Virtual Life: The Culture and Politics of Information* (San Francisco: City Lights, 1996); Scott Bukatman, *Terminal Identity: The Virtual Subject in Postmodern Science Fiction* (Durham: Duke University Press, 1993); Manuel Castells, *The Informational City: Information Technology, Economic Restructuring, and the Urban-Regional Process* (Oxford: Blackwell, 1989); Manuel Castells, *The Rise of the Network Society*, vol. 1 (Oxford: Blackwell, 1996); Verena Andermatt Conley, ed., *Rethinking Technologies* (Minneapolis: University of Minnesota Press, 1993); Jonathan Crary and Sanford Kwinter, eds., *Zone 6: Incorporations* (New York: Zone Books, 1992); Mark Dery, ed., *Flame Wars: The Discourse of Cyberculture* (Durham: Duke University Press, 1994); Gary Downey and Joe Dumit, eds., *Cyborgs and Citadels: Anthropological Interventions in Emerging Sciences and Technologies* (Santa Fe: School of American Research Press, 1997); Donna Haraway, *Simians, Cyborgs and Women* (New York: Routledge, 1991); Donna Haraway, *Modest_Witness@Second_Millennium. FemaleMan©_Meets_OncoMouse™: Feminism and Technoscience* (New York: Routledge, 1997); David Hess, *Science and Technology in a Multicultural World* (New York: Columbia University Press, 1995); Donald MacKenzie and Judy Wajcman, eds., *The Social Shaping of Technology* (Philadelphia: Open University Press, 1985); Aihwa Ong, *Spirits of Resistance and Capitalist Discipline: Factory Women in Malaysia* (Albany: State University of New York Press, 1987); Constance Penley and Andrew Ross, eds., *Technoculture* (Minneapolis: University of Minnesota Press, 1991); Howard Rheingold, *Virtual Reality* (New York: Simon and Schuster, 1992); Tricia Rose, *Black Noise: Rap Music and Black Culture in Contemporary America* (Hanover: Wesleyan University Press, 1994); Andrew Ross, *Strange Weather: Culture, Science and Technology in the Age of Limits* (New York: Verso, 1991); Richard Sclove, *Democracy and Technology* (New York: Guilford, 1995); Clifford Stoll, *Silicon Snake Oil: Second Thoughts on the Infor-*

mation Highway (New York: Doubleday, 1995); Sherry Turkle, *The Second Self: Computers and the Human Spirit* (New York: Simon and Schuster, 1984); Sherry Turkle, *Life on the Screen: Identity in the Age of the Internet* (New York: Simon and Schuster, 1995); Langdon Winner, *The Whale and the Reactor: A Search for Limits in an Age of High Technology* (Chicago: University of Chicago Press, 1986).

2. Herbert Schiller, *Information Inequality: The Deepening Social Crisis in America* (New York: Routledge, 1996); Herbert Schiller, *Information and the Crisis Economy* (Norwood, NJ: Ablex, 1984); William Wresch, *Disconnected: Haves and Have-Nots in the Information Age* (New Brunswick: Rutgers University Press, 1996); Stoll, *Silicon Snake Oil.*

3. Stanley Aronowitz and William DiFazio, *The Jobless Future: Sci-Tech and the Dogma of Work* (Minneapolis: University of Minnesota Press, 1994). For other discussions of class, labor, and technology, see Stanley Aronowitz and Jonathan Cutler, eds., *Post-Work: The Wages of Cybernation* (New York: Routledge, 1998); Castells, *The Rise of the Network Society;* Barbara Garson, *The Electronic Sweatshop* (New York: Simon and Schuster, 1988); Ong, *Spirits of Resistance and Capitalist Discipline;* Andrew Ross, "Jobs in Cyberspace," in *Real Love: In Pursuit of Cultural Justice* (New York: New York University Press, 1998), 7–34.

4. Castells, *The Rise of the Network Society.*

5. For discussions of gender, feminism, and technology, see Ann Balsamo, *Technologies of the Gendered Body: Reading Cyborg Women* (Durham: Duke University Press, 1996); Berner, Boel, ed., *Gendered Practices: Feminist Studies of Technology and Society* (Linkoping, Sweden: Department of Technology and Social Change, Linkoping University, 1997); H. Patricia Hynes, ed., *Reconstructing Babylon: Essays on Women and Technology* (Bloomington: Indiana University Press, 1991); Gill Kirkup and Laurie Smith Keller, eds., *Inventing Women: Science, Technology, and Gender* (Cambridge: Blackwell, 1992); Cheris Kramarae, ed., *Technology and Women's Voices: Keeping in Touch* (New York: Routledge and Keegan Paul, 1988); Sadie Plant, *Zeros and Ones: Digital Women + the New Technoculture* (New York: Doubleday, 1997); Jennifer Terry and Melodie Calvert, eds., *Processed Lives: Gender and Technology in Everyday Life* (New York: Routledge, 1997); Joan Rothschild, *Machina ex Dea: Feminist Perspectives on Technology* (New York: Pergamon Press, 1983); Judy Wajcman, *Feminism Confronts Technology* (University Park: Pennsylvania State University Press, 1991).

6. Plant, *Zeros and Ones.*

7. For example, the edited collection *Wired Women: Gender and New Realities in Cyberspace* begins with the following disclaimer, "None of the few women of color we were able to find online were available to write, a mirror of the extremely white nature of the medium at this time." They unwittingly assume

the female version of the white geek icon even as they try to undo its implicit maleness. See Lynn Cherny and Elizabeth Reba Weise, eds., *Wired Women: Gender and New Realities in Cyberspace* (Seattle: Seal Press, 1996).

8. Scientific racism and eugenics have been supported in part by technology's ability to "mark" racial differences. Further, as Michael Adas argues, the presumed lack of technologies in certain parts of the world, including Africa and India, have been used to support a theory of racial inferiority and as a justification for colonial domination. See Michael Adas, *Machines as the Measure of Men: Science, Technology, and Ideologies of Western Dominance* (Ithaca: Cornell University Press, 1989). On surveillance and information gathering, see Mike Davis, *City of Quartz: Excavating the Future in Los Angeles* (New York: Vintage, 1992); James Lyon, *The Electronic Eye: The Rise of Surveillance Society* (Minneapolis: University of Minnesota Press, 1994); and Oscar Gandy, Jr., "The Panoptic Sort," in Brook and Boale, *Resisting the Virtual Life*.

9. See Michel de Certeau, *The Practice of Everyday Life* (Berkeley: University of California Press, 1984); Henri Lefebvre, *The Critique of Everyday Life* (London: Verso, 1991); and Kristin Ross, *Fast Cars, Clean Bodies: Decolonization and the Reordering of French Culture* (Cambridge: MIT Press, 1996).

10. As Lefebvre points out, the emphasis on work and productivity that has marked the modern world is ironically reinforced by Marxist theorists who see any struggle outside the class struggle as irrelevant:

> By starting from an abstract notion of the class struggle, some Marxists have neglected not only to study the recent modifications of capitalism as such, but also the "socialization of production," and the new contents of specifically capitalist relations. Such a study could perhaps have modified the notion of class struggle, leading to the discovery of new forms of struggle.

Lefebvre, *Critique of Everyday Life*, 38.

11. Ibid., 32–33.

12. See Penley and Ross, *Technoculture* for other examples of technological "refunctioning." We borrow the concept "cultural priority" from Tricia Rose; for further elaboration, see Rose, *Black Noise*, chap. 3.

13. *BJ and the Bear* was a popular television show about a trucker and his chimpanzee.

Chapter One

Beyond Access

Race, Technology, Community

Logan Hill

When Microsoft's founder Bill Gates and Vice President Al Gore both appear in Harlem to talk about the "digital divide," most people cringe. Most assume that the bombastic rhetoric of such leaders is just politics as usual or the latest stab at niche marketing, and much of the time they're right. But in a quickly changing digital era of quicksilver developments, deals, and innovations, the racial disparities in computer access, training, and employment have been lamentable constants. Despite pervasive promises of a glorious digital democracy, we cannot afford to simply brush off the digital divide. We must confront it.

Even as IPO-hungry companies herald the uncertain ascendance of free or nearly free PCs, federal studies have found that the digital divide continues to widen.[1] Minorities are still two to three times more likely to lack a simple telephone, the fundamental bridge from American homes to the Internet.[2] African American and Latino households are about half as likely to own a computer at home than whites and Asians (computer penetration in 1998: Asian/Pacific Islander, 55 percent; white, 46.6 percent; Hispanic, 25.5 percent; black, 23.2 percent), and as much as three times less likely to have Internet access.[3] Schools, libraries, and businesses in predominantly minority communities are all much less likely to have high-speed information services. Even if they do have access, it probably isn't as fast as many white consumers', they probably don't have as much training, and they're less likely to be working in high-paying technology jobs. And according to the most recent federal study of the digital divide, the gap between the information haves (the rich, white, Asian, and urban-dwelling populations) and have-nots (the poor, Hispanic, black, American Indian, or rural-dwelling populations) is growing.[4]

Unfortunately, these basic facts about minority communities and digital culture have been obscured in a deluge of Wall Street–addled rhetoric and press releases. Online companies have exaggerated ethnic participation on the Web in order to preserve the ideal of diversity so essential to their advertising pitches and market valuations. In practice, however, the democratic promises of the digital revolution remain as unfulfilled as the rest of our civil rights dreams. But there is hope.

After surveying this century's technological transformation in the epochal Information Age trilogy, Manuel Castells looked toward this new century in 1998, and argued that a "new politics will have to emerge":

> This will be a cultural politics that starts from the premise that informational politics is predominantly enacted in the space of media, and fights with symbols, yet connects to values and issues that spring from people's life experience in the Information Age.[5]

This new digital politics has already begun to emerge. Of course, it is as fractured and fraught with contradictions as any progressive political movement. In the United States, it has manifested itself in four primary sites (and, as the other essays in this anthology stress, many more that aren't so obvious):[6]

1. High-tech labor. Most digital CEOs are white; most digital workers are not, even here in the United States (globally, the problem is much more severe). For years, activists have been organizing high-tech workers to insure that the factory work, tech support, lower-level programming, and manual work that make the digital industries run are not ignored. Organizers have successfully lobbied for better working conditions, jobs, pay, training, and benefits for minorities in the tech industries.
2. Universal access. Activists and legislators have long fought for low-cost service and subsidies that fund the equitable wiring of schools, libraries, and rural, low-income, or minority communities.
3. Community technology centers (CTC). Often as part of schools,

public libraries, or housing developments, these local centers offer immediate information technology access and training.

4. Racial/ethnic content providers. This new wave of digital designers, firms, and companies largely blame white-bread content for the disparities in computer access and training. In response, they sustain sites, companies, and services for underrepresented communities, in order to bring more minorities online, and often to profit from their emerging markets.[7]

The digital divide isn't just about personal computers; it's about training, access, education, content, telecommunications infrastructure, and more. These four sites have been chosen because they tackle the divide from very different—even adverse—perspectives. They don't always work together, but they complement each other nonetheless.

The high-tech labor movement evolved out of a long history of labor activism, one that has increasingly learned to adapt to the new information economy. Because their progress has been documented by writers like Stanley Aronowitz and Aihwa Ong, this essay will focus on the three relatively less often discussed sites: universal access, community technology centers, and ethnic content providers.[8] The universal access debate remains the chief public battleground over online civil rights in public discourse; community technology centers work on the ground and face-to-face with local communities; ethnic content providers most often arrive at the train-wreck collision of symbolic politics and material profits.

While rhetorical wars rage, these three movements have consistently yielded tangible rewards. Access advocates have lobbied and legislated to wire thousands of homes and schools. CTCs have proven that poor communities need—and prosper from—the presence of hardware and training immediately. Racial/ethnic content providers attempt to offer targeted information, entertainment, services, and commerce in order to bring underrepresented communities online. Each faces different practical racial dilemmas every day, whether it's in the pages of public opinion, in the halls of Capitol Hill and Silicon Valley, or in disenfranchised communities. But what they share in common is the belief that technologies are a very real site of political struggle.

Universal Access

Digital divide advocacy's most direct historical precedent is the universal access movement, which followed the invention of the telephone. Alexander Graham Bell's device was first regarded as a curiosity, a novel, if miraculous, toy with little practical value that would be enjoyed only by wealthy dilettantes and tinkering hobbyists. However, as the phone's revolutionary convenience—economic and otherwise—became more apparent, the Federal Communications Commission and others argued that to refuse phone access in rural or low-income areas was to deny a basic utility. These advocates argued that telephone service, like water or light, would become a fundamental part of the American infrastructure and everyday life. Without inexpensive phone service, rural businesses and residents would suffer especially, since telecom companies were unlikely to spend money wiring the least profitable rural regions of the country. Liberal legislators employed populist rhetoric to tie telephone access to democracy, while conservative politicians derided their innovative efforts as the next frivolous, bells-and-whistles expense of the liberal Roosevelt bureaucracy. Sound familiar?

It does to the first African American director of the National Telecommunications and Information Administration (NTIA), Larry Irving, who spearheaded the government's "Falling through the Net" studies, the primary and most reliable studies of computer use and Internet access in the nation. Irving conducted his studies during the Clinton administration's tenure, and the data his department collected were used to support Al Gore's massive Education Rate (E-Rate) initiative to wire all American schools. All of Irving's work was consciously conducted under the auspices of "universal access," a precedent-setting definition first used by the FCC in the 1940s to insure that every American household would have affordable access to telephones.

The NTIA's first substantial attempt to document the growing digital information gap was the 1995 study "Falling through the Net: A Survey of the 'Have Nots' in Rural and Urban America." The study first reminded the country that a gap persisted in telephone access. Rural and central city residents had substantially less access to simple telephones than other groups. Though 93.8 percent percent of American households had telephones, the study found that Native Americans in rural areas have the fewest telephones (75.5 percent), followed by rural Hispanics (79 percent) and rural African Americans (80.9 percent). By contrast, 95.4 percent of white rural households were found to have telephones. This was nothing new. What was news was the fact that "black households in central cities and particularly rural areas" owned the lowest number of PCs, with rural Hispanics close behind.[9] The study found that white households were about three times more likely than African Americans and Hispanics to own computers.[10]

Interestingly, the study found that low-income users ($10,00–$14,000) were "the most likely users of online classes," such as correspondence courses. Minorities were also found to search classifieds and access government reports more frequently than whites (and subsequent studies have shown that minority computer users are more likely to use the Web to navigate civic bureaucracies). The study asserted that those who had the least access to computers needed them most and even used their most vaunted applications—a fact that has been reiterated in each subsequent federal study. In conclusion, this 1995 study presciently called for public schools, libraries, and "community access centers . . . to complement a long-term strategy of hooking up all those households who want to be connected."[11] That recommended strategy included tech education programs, community technology centers, pro-competition legislation, and legislation incorporating tax breaks and other price breaks for schools and libraries as part of a larger effort to make information access affordable for everyone. The study's pointed recommendations composed the most forceful declaration of a revived universal access movement.

That groundbreaking new data led to a subsequent, more thorough study in 1998, "Falling through the Net II: New Data on the Digital Divide." The product of more than forty-eight-thousand door-to-door surveys in 1997, the study concluded that the divide had become more severe, not less, since 1995: "There is a widening gap, for example, between those at upper and lower income levels. Additionally, even though all racial groups now own more computers than they did in 1994, Blacks and Hispanics now lag even further behind Whites in their levels of PC-ownership and on-line access."[12]

Again, the July 1999 study, "Falling through the Net: Defining the Digital Divide," found that this gap had continued to expand into what the government then described as a "racial ravine." While the federal studies confirmed that a racial gap existed, they could not establish more than correlations for why that gap had appeared and widened. So in the mid-1990s opinion makers—from Marxists to cyberculture of poverty conservatives—began to debate the causes and implications of a digital divide.

The most explicit site of this debate was the ambitious E-Rate initiative proposed by President Clinton and Vice President Al Gore. The E-Rate program, as Gore presented it in 1996, would subsidize more equitable access by charging a long-distance telephone bill of about one dollar a month, and use those funds to wire schools. At last, Democrats found that access was a racial issue they could finally sink their teeth into. From an administration that backed away from welfare and affirmative action, and whose largest racial initiative was the sponsorship of "dialogue," the sight of the vice president citing hard disparity statistics and pushing for the E-Rate's substantive legislation was a welcome departure. Clinton promised to put "a computer in every classroom by the year 2000"—a promise that was ultimately not fulfilled—and one in every home by 2007, so that "every twelve year old can log onto the Internet."[13]

But as Internet traffic soared in the nineties and influential cyber-boosters announced that universal access was inevitable, racial disparities began to look less worrisome. Libertarian resistance to government intervention began to dominate political discussion in the tech industry. Conservatives, libertarians, and reformist neo-Luddites have since used antitax populism to portray universal technology access as just another expensive, misguided liberal program; on the other hand, anticorporate leftists like Herbert Schiller have derided high-tech programs as insidious corporate conspiracies.[14] The E-Rate plan was condemned as a simple "Gore Tax" in 1998; its budget was sliced the first year of implementation, from $2.5 billion to nearly half that. As it stands, the program has wired approximately eighty thousand schools, but its future remains uncertain.[15]

At its worst, the digital disparity was lampooned as a hoax. The syndicated Americans for Tax Reform columnist James Glassman, for example, argued that politicians like Gore support computer access because it's popular in the polls and with powerful tech lobbyists (though the tech industry has hardly been supportive of subsidies): "The political benefits to Gore of linking schools, high-tech and the doling-out of billions of dollars in cash seemed obvious. The educational benefits are more uncertain, and 80 percent of schools are already connected to the Internet anyway."[16] Glassman's argument ignores the fact that federal studies in 1998 found that fewer than 30 percent of schools had even one classroom wired to the Internet (most had isolated computer labs), and the ones not wired are almost all located in low-income and predominantly minority areas.[17] A 1998 study by the Department of Education's National Center for Education found that schools with 50 percent or more minority students had Internet access in only 5 percent of their classrooms, compared to 18 percent in schools where less than 20 percent of the student body was composed of minorities.[18] Race and class cannot be removed from the access equation.

Still other critics have sought to distract public attention from technological inequalities with arguments for a return to the three Rs. Neo-Luddite educational critics like William L, Rukeyser of Learning in the Real World have interpreted the arguments of access and technology education advocates as evidence of "a tremendous faith that computers were in fact some plaster saint that would save the day."[19] His organization seeks to debunk the miraculous claims of computers' effectiveness, and focuses on the supposedly terrible travails of ultra-privileged, bleary-eyed, carpal-tunnel preschoolers who've forgotten what it's like to play in the sand.

These reactionary arguments are partially the fallout of the information industry's own hyperbolic pronouncements, which have, of course, often been ridiculous. Notably, Steve Jobs backed away from his industry boosterism and told *Wired*, "I used to think that technology could help education. I've probably spearheaded giving away more computer equipment than anyone else on the planet. But I've come to the inevitable conclusion that the problem is not one that technology can hope to solve."[20] Yes, smaller class sizes and higher teacher pay might certainly do more to improve schools than a blueberry iMac, but it doesn't have to be one or the other; there's no reason to give up on information technology. Unlike Jobs, Gregg B. Betheil, an assistant principal at Martin Luther King Jr. High School in New York, never expected that dropping computers in schools would "solve" educational problems. When a similarly dismissive argument appeared on the editorial page of the *New York Times* in July 1999, Betheil responded.

> The divide is not about wires. It is about access—to information, knowledge, culture and to that side of society that has always had more. To call for expanded access yet say "it may be time to throw in some new books, too" simply misses the point. The books should have been there all along. Once again, the poor must wait for hand-me-downs from unhappy, private elementary-school students who don't like their laptops.[21]

These sorts of systemic information inequalities will persist unless we take action on universal access and insure that digital services are available and affordable in all communities.

In the dot-com giddiness of the nineties, we've come to believe the glowing promises of Nasdaq without looking pragmatically at the future, or even at the underprivileged in our own communities. In order to insure that the legacy of the e-commerce boom is more than profits for a few, we must take seriously the NTIA's recommended course: pro-competition telecommunications policies, support for CTCs, and policies that will reduce information access costs, particularly for schools and libraries. Like it or not, the government may be the only body capable of negotiating with the formidable telecommunications and digital industries to guarantee universal information access. But as advocates continue to struggle for these types of macro-level reforms, thousands of activists are working for more temporary, local remedies.

Community Technology Centers

Rather than wishing on campaign promises and depending on corporate charity, community technology centers are compensating for a history of denied access on the ground. The members of organizations like Playing to Win (Harlem), Street-Level Youth Media (Chicago), Plugged In (East Palo Alto), Break Away Technologies (Los Angeles), the MIT Computer Clubhouse (Cambridge), and the National Urban League are actually bringing information technology to low-income and minority schools, libraries, businesses, and communities. Amid high-flung theoretical posturing and the flame wars of radical online communities and listservs, CTCs have quietly wired up hundreds of thousands of Americans.

"There is a nationwide movement gathering steam for community empowerment via technology," said Phil Shapiro, a director of CTCNet, an association of more than three hundred CTCs across the United States. "It's no longer a fringe kind of movement."[22] Following in the footsteps of prescient groups like the National Urban League, which began training African Americans on IBM mainframes as early as 1976, CTCs have been springing up across the nation in the last decade.

Few community groups have been as successful as Harlem's local Playing to Win, founded in 1980 by Antonia Stone and still running today. Stone, who left the organization to found CTCNet, wrote a brief in-house history of Playing to Win that stresses the basics:

> Recognizing that in our increasingly technological society, people who are socially and economically disadvantaged will become even further disadvantaged if they lack access to computers and computer-based technologies, Playing to Win was founded in 1980 to promote and provide access to computer use and technology education for those who typically would lack such opportunity. Playing to Win's emphasis on the development of technological capability as a route to achieving personal goals and empowerment has remained constant, but the need for such an organization has increased dramatically as technology, in the intervening years, has pervaded every aspect of work and learning in our society.[23]

Community technology centers like Playing to Win do not, as some mistakenly and optimistically believe, help minorities get ahead; they help them keep up. CTCs are not get-rich-quick schemes for racial uplift, but a reaction to inaction and the awful disparities documented by studies, but more importantly, lived by millions of individuals every day.

Centers like Playing to Win are distinguished by their simplicity. Just a block away from a low-income housing development, the volunteers of Playing to Win work in a cinderblock space that might well have been a corner store. The space is small, but crammed full with computers. A separate lab allows the young HarlemLive Web site members (mostly high school students) to write and produce their stories on the Web and to, in their words, "showcase [their] talents while learning about emerging technologies." While HarlemLive is geared toward teenagers, Playing to Win serves users of different ages and backgrounds. First-time users of all ages learn basic computing skills from email to surfing, while intermediate students experiment with digital design, production, and coding. Centers like these facilitate job hunting, training, and the simple sorts of play and casual exploration that the information-rich take for granted.[24]

Casual computer use, those less obvious times when computer users find access outside the home, school, and office, enables a level of comfort with technologies that is crucial to active computing. Several recent studies have illustrated the ease with which white or Asian students find computers at friends' homes, or the libraries and community centers near their homes, compared to other students who live in less affluent or less computer-savvy neighborhoods. In their 1998 study "Bridging the Digital Divide," Donna L. Hoffman and Thomas P. Novak of Vanderbilt University found that nearly 38 percent of those white students who do not own a computer at home still reported using the Internet in the past six months, as opposed to 16 percent of the African American students who do not own computers. White students benefit from technologically rich social networks that minority students lack. CTCs are just one way to give minorities an artificial network that can ameliorate their disadvantage.[25]

Because CTCs deal with individuals and not numbers, Ramon Harris, who directs the Executive Leadership Foundation's Transfer Technology Project, a program that works to enhance computer training at traditionally black colleges, says statistical studies of the divide have been too simplistic. "Of course there's more to computer usage than economics," he says simply, "you have to take a look at the people you're talking about." Harris explains,

> Let's say you may make good money as a steelworker or a construction worker. Your children are less likely to grow up using computers, but not because your income is bad. A lot of blue-collar employees are solidly middle-class. It's because you're not likely to use computers at work; it's because there's nothing online that you need or desire, so you don't encourage their use at home. Skilled workers are not predisposed to technology and if you're not connected at home, studies have shown that you're less likely to use information technology regularly.

As Harris notes, most studies (including the federal "Digital Divide" series) haven't looked at more than simple income statistics, and have failed to note that minorities are just not well represented in white-collar jobs. Says Harris, "If you don't have the same access [in grades] K through 12, then you don't move into college at the same rate, or

[into] the high-tech, financially rewarding jobs."[26] He works to distribute technology to minority communities because he believes that such access will help reduce not just the digital divide, but the severe income gap in the United States.

Similarly, the National Urban League's B. Keith Fulton says researchers must consider other factors that don't show up in simple access studies, including the possibility that minority Americans may not use online content or apply for online jobs as frequently because of a lack of enticing content or because of racism in the industry. A major challenge facing CTCs today is how to encourage minority participation in a digital culture that is dominated by white producers and consumers. As a result, CTCs must temper attempts to offer material access and practical training with a sensitive understanding of their community's needs and desires. For, as Fulton has noted, "Computer ownership will increase when there is clear reason for it. Not your reason or my reason, but clear for the potential owner."[27]

The real role of CTCs, says the former director of CTCNet, Peter Miller, goes far beyond simply supplying machines.

> The development of neighborhood centers is not only an economically practical strategy on the road towards universal access, it's an eminently sensible social and political approach, too, strengthening and building local institutions which are the foundation of self-help and empowerment.[28]

The brilliant advantage of CTCs is that they can offer valuable, essential services and training that wouldn't otherwise exist, and boost user confidence and training along the way. At Harlem's Playing to Win center, one teenager is already working as a systems administrator for a local company, another is designing Web sites for uptown rap and hip hop labels, and several are interning with emerging dot-coms. They learn basic skills quickly, and Playing to Win supplies them with the regular professional contacts that allow them to put those skills to use. As Antonia Stone sees it, "Technology access is not an end in itself, but rather a means to educational, social, political, and economic opportunity."[29]

But as CTCs proliferate, they face their own digital divides. Centers in relatively wealthy, information-rich areas are thriving, but others in more rural areas are not. Indeed, many of these communities are overlooked, especially when companies decide to make bold public relations moves and head to media-saturated locales like Harlem or Compton.[30] As Susan Myrland, a San Diego community technology consultant, has argued, high-profile regions like New York or Los Angeles have untold benefits, since they "regularly receiv[e] corporate support and news coverage." According to Myrland, San Diego is often overlooked; tech centers in Los Angeles and San Francisco take the bounty of funding.[31] This produces competition among CTCs hungry for limited funding; those in rural areas, where they are perhaps needed most, may be the hardest hit. And if the e-business boom turns out to be just a burst, CTCs could be devastated by the fallout of tighter company budgets.

The only way to sustain CTCs over the long term may be through networks. David Geilhufe, the executive director of Oakland's Eastmont Computer Center, says, "We estimate that there are about 3,000 total centers around the country." He adds, "It makes a lot more sense to link up those operations," because digital equity "won't happen because one organization has a center, but because 100 organizations have a network of centers."[32] Networks of centers are thriving, particularly those sponsored by the government. The U.S. Department of Housing and Urban Development (HUD) has launched almost 500 technology centers in the last four years, and plans to build 760 more in the near future. Meanwhile, the Department of Education has promised to build more than six hundred centers. Norris Dickard, the director of the department's community technology program, says optimistically, "We will blanket all of our low-income communities."[33]

Community technology centers are very young institutions, so an evaluation of their long-term success is nearly impossible. But it's also indisputable that they've moved thousands of young people into high-tech jobs, who might otherwise have not had the opportunity. Still, regardless of their efficacy, we must remember that CTCs cannot possibly solve the problem alone. As the 1999 "Falling through the Net" study states, "Whites are more likely to have access to the Internet from home than Blacks or Hispanics have from any location."[34]

Racial/Ethnic Content Providers

In April 1999, the Rainbow Coalition invested a symbolic $100,000 in fifty Silicon Valley corporations in order to allow minorities access to shareholder meetings, as part of what Jesse Jackson has described as the fourth stage of the civil rights movement: "the right to capital."[35] The group's high-profile investment provoked a national discussion about the paucity of African Americans in the information industries and the lack of minority-friendly content available online. Indeed, of the 384 directors of Silicon Valley's largest fifty firms, only five are black and one is Latino.[36] English is the lingua franca of the Web, as experts estimate that around 80 percent of the content on the Web is posted in English, though more than 40 percent speak another language every day.[37] Major Web magazines like *Slate* and *Salon* are as white as the *New Yorker* before Hilton Als and Henry Louis Gates showed up. These conditions make cyberspace less than enticing for minorities.

The racial/ethnic content deficiency has been obvious, but solutions have not. Web content is mainly funded by online sales or advertising (nonpornographic subscription sites are doing very poorly), and studies that demonstrate low minority and low-income participation are used as arguments against funding ethnic-specific content, just as they were deployed in radio, film, and television. In fact, on TV, minority programming has been fostered not by media giants like ABC or CBS, but by ethnic-specific companies like BET and Telemundo. Similarly, ethnic-specific companies may be the Web's best hope for diverse content.

The chairman of the Los Angeles–based AfroNet Web site, Willie Atterberry, says his product "allows minorities to come into an area that's populated by themselves and learn how to use the technology. . . . You don't have to worry about going to Silicon Valley. It comes to you."[38] Atterberry and other racial/ethnic content providers argue that their sites, portals, and networks are the solutions to low minority participation. Their content is intended to bring in minority computer users to a white-dominated online world, turn a profit, and transform the Internet in the process.

In the past few years, Web sites dedicated to serving specific racial and ethnic communities have proliferated as more minorities got online and the need—or market demand—grew stronger. BET has created a black online presence, StarMedia and Yupi opened up for Latinos, sites like RedRoad or SeminoleTribe courted American Indians, and AsianAvenue has catered to Asians. Of course, thousands of noncommercial Web zines, sites, listservs, and chat rooms have fostered ethnic communities online, but in the last few years, commercial sites have had perhaps the greatest impact on online minority participation. Sites like Philippine News Link and AfricaNews provide in-depth news or links to news that mainstream sites wouldn't carry, and supplement their coverage with chats for interested surfers. Other sites like LaMusica provide expert information and Webcasts of specific musical genres that might not get airplay on mainstream media.

Racial/ethnic content sites are crucial because they give minorities reasons to go online. Without something rewarding, whether it's that live song by a favorite musician, news about a family far away, or a simple catalog in another language, casual computer users would have few personal reasons to join the information economy, whatever their ethnicity. Ethnic content providers simply offer all the services, news, chats, discussions, organizations, and resources that mainstream sites offer, but with a more particular focus. These sites allow new computer users to see something of themselves in an unfamiliar place, but they also facilitate the sorts of networks, professional and otherwise, that so often lead to friendships, community organizations, political change, and commerce.

Each content provider presents information specifically tailored to at least one ethnic community, and each has fought obstacles and stereotypes that mainstream sites rarely face. Though the experience of Fernando Espuelas, the multimillionaire CEO and president of the Latin American portal StarMedia Inc., is exceptional, it is certainly an instructive case study. StarMedia's goal was to provide both locally specific and general content for Latin Americans throughout the Americas. "To reach our audience we have to be transnational and national at the same time," the thirty-something Espuelas says, "because our audience is actually very diverse." StarMedia set up offices in countries like Argentina, Brazil, and Mexico to produce content in Spanish, Portuguese, and even in regional dialects. As Espuelas says, "If you go into StarMedia Mexico you'll find a different kind of Spanish. But underlying that is the sense that an Argentine and a Mexican can communicate, and not only can, but want to communicate."

Such communication is more important than ever now that minorities appear in relatively small numbers online. Internet portals like StarMedia make it easier for Latinos to find each other and to share information relevant to their own experiences. This is not to say that minority computer users want to isolate themselves from the rest of the Internet, but it does reflect the fact that many of the most successful Internet sites are niche pages that allow users to go deeper into their own particular interests than they could in traditional media.

Because Espuelas's business model was so unusual, it was greeted with cynicism and misunderstandings on Wall Street. "We had to sell ourselves by selling Latin America," he says, because he found that "most investors still have this Carmen Miranda view of Latin America—lots of happy people dancing with fruit on their heads—or one of those 1960s westerns, big desert, and a dusty town full of people in sombreros waiting for the coup to come." Though companies like StarMedia are now hot tech properties, Espuelas admits, "we literally came very close to going out of business because we spoke to Internet investors who didn't understand Latin America at all. I mean, we had people asking, Do they have phones?"[39]

Still, CEOs like Espuelas are not flocking to the Internet out of selfless determination. They're looking to make a buck while providing a service, which places such businesses in a very tricky position. Though Espuelas fervently believes that "this technology will have an exponentially greater effect on disempowered people and will drastically improve the lives of Latin Americans," he also admits, "I'm a cold-hearted capitalist." He opposes government intervention and critiques universal access plans from a purely economic stance. "How do you sustain this," he asks, "without making a Soviet Union out of the Internet? Someone has to pay for these programs."[40] While private businesses may indeed provide enticing reasons for minority computer users to jump online, we should not look for them to serve users' interests only; their bosses are, after all, not just minority computer users, but shareholders.

Corporate racial/ethnic content providers walk the fine line between commodifying a culture and serving it every day, but unfortunately, the Internet seems to be no more hospitable to nonprofit enterprises than radio or television, particularly as content becomes more broadband and expensive to produce. Already, there are currently no nonprofit Latino Web sites that can compete with the deep content of a StarMedia. Though the Internet was heralded as a low-cost platform for absolutely anyone, online services have now matured, and it has become more difficult to produce high-quality, thorough content without a large investment. As more and more computer users demand broadband multimedia content, it is becoming clear that content providers cannot cover their costs without a large and diverse demographic. This could be difficult for small minority sites, but it could be positive in the long run for minority representation.

The fact is, the Internet needs minorities as much as minorities need the Internet. Bill Gates explained,

> The Internet won't attract enough great content to thrive if only the most affluent 10 percent of a society can get to it. There are fixed costs to authoring content, and a large audience is required to make it affordable. In the long run the broadband information network is a mass phenomenon, or it is nothing.[41]

E-commerce needs minority participation online if it is to survive this giddy stock boom, particularly as the demographics of the United States continue to shift. But minorities also need content that speaks to their needs and communities. Sustainable, diverse sites geared toward minorities may well determine the future ethnic composition of the Internet, if not the fate of the Internet itself.

Conclusion

Techies often repeat that talent—good code—is all that matters online, not race, social connections, or capital. In a recent *Silicon Alley Reporter* roundtable discussion on access, Doubleclick CFO Stephen Collins asked, "What's the problem? I have seen a toaster in the Williams Sonoma catalog that was more expensive than a computer."[42] While a chief beneficiary of one of the Internet's most astounding IPOs may claim that he doesn't "see anything happening that's exclusionary," some disagree. "Access is still in a consumer mentality," says Peter Miller, a former director of CTCNet. "It's compared to swimming pools or tennis courts," and not the systemic discrepancies evident in education or job training. Computer access is not a luxury, it's a necessity. To stress the importance of this digital divide, Tony Riddle, the executive director of Manhattan Neighborbood Network, often reminds audiences of Juneteenth, the day in 1865 when slaves in Texas learned about the Emancipation Proclamation—two and a half years after the fact. The United States has a clear history of racialized information gaps.[43]

Yes, soon the "digital divide" as we know it will vanish. Nearly everyone will be online—when they watch "television," listen to music, buy groceries, or go to the bank. But a divide will still exist. No matter what promises the technoptimists make, it seems certain that under supposedly "friction-free" market conditions, rich people living in communities of knowledge workers will continue to access more information at faster speeds.[44] New, high-speed Internet access has always gravitated toward consumers with the most disposable income. Without professional access to means of production (high-speed connections, software, and hardware), the users of passive technology like WebTV are little more than consumers in an information economy. Universal access isn't just about being able to surf the Web, it's about the ability to participate and compete in a technology-driven industry and society. Information-poor communities not only miss out on interesting content, they do business at a severe disadvantage, in what Manuel Castells describes as "informational black holes": information goes in, but it doesn't come out.[45]

As the testimonies in this essay reveal, the digital divide is not just a technical problem of how to make the Web faster, easier, and more user-friendly, or even simply a question of how to get more computers to more people. Though undoubtedly important, hardware accessibility does not solve the problem of how to more evenly distribute technological resources and knowledge, including skills training and job opportunities. It has become clear that the complexities of the digital divide force us to think about access in much broader terms, as a challenge that requires multiple avenues of intervention. Universal access advocates and legislators fight for policy changes as CTCs wire and train thousands of underprivileged Americans. Ethnic content providers are giving minorities a reason to log on, and betting money that they will flock to the Web in numbers. Collectively they provide some important, if imperfect and incomplete, steps to narrowing the gap. As Ramon Harris says, "Technology is an investment. Either you invest now and insure that there's access and availability, or you insure that there will be an underclass."[46]

NOTES

1. "Falling through the Net II: New Data on the Digital Divide" (Washington, D.C.: National Telecommunications and Information Administration, 1998).

2. High-speed services (cable modems, T-1 or T-3 lines, DSL, etc.) are conspicuously unavailable in most minority communities. American Indians have consistently shown the lowest rates of telephone availability.

3. "Falling through the Net: Defining the Digital Divide" (Washington, D.C.: National Telecommunications and Information Administration, 1999).

4. Ibid.

5. Manuel Castells, *End of Millennium* (Oxford: Blackwell, 1998), 352–53. According to Castells, "The rise of informationalism in this end of millennium is intertwined with rising inequality and social exclusion throughout the world" (70). He and others have illustrated that the digital divide is most severe as a global problem. In this essay, I concentrate on the terrible internal divide in the United States, the most technologically privileged country in the world.

6. Castells himself has articulated several ideas for future revolutions in information access in the article "The Informational City Is a Dual City: Can It Be Reversed?" His proposals include greater information and educational access for prison inmates, augmented high-tech education, telecommuting centers, community media, and a renewed commitment to local governments. In Donald A. Schön, Bish Sanyal, and William J. Mitchell, eds., *High Technology and Low-Income Communities: Prospects for the Positive Use of Advanced Information Technology* (Cambridge: MIT Press, 1999), 25–43. Certainly, digital education reformers are critical to training and educating minority Americans and closing the divide.

7. I consider these four sites to be the most crucial. By no means are they the only sites. One could argue, for instance, that the antimonopoly cadres encourage the redistribution of wealth and information power from centralized, predominantly white institutions to smaller companies. Andrew Shapiro's book *The Control Revolution* (New York: New Press, 1999) stresses many of the ways activist Web sites can galvanize and improve communities.

8. For an introduction to high-tech labor, see Stanley Aronowitz and William DiFazio, *The Jobless Future: Sci-Tech and the Dogma of Work* (Minneapolis: University of Minnesota Press, 1994); Stanley Aronowitz and Jonathan Cutler, eds., *Post-Work: The Wages of Cybernation* (New York: Routledge, 1998). For information on women and high-tech labor, see Aihwa Ong's groundbreaking *Spirits of Resistance and Capitalist Discipline: Factory Women in Malaysia* (Albany: State University of New York Press, 1987).

9. "Falling through the Net: A Survey of the 'Have Nots' in Rural and Urban America" (Washington, D.C.: National Telecommunications and Information Administration, 1995), 1–3.

10. The federal studies have understood access mainly in terms of simple home computer ownership, but many have argued against the inclusion of stripped-down services like WebTV. In William Mitchell's article "Equitable

Access to the Online World," he writes, "Connections should also be two-way, and . . . symmetrical. One-way connections, like those established by broadcast media, create a rigid division between producers and consumers of information" (153). Regardless, as consumers expressed their desire for greater control over information as content creators, the NTIA reported that passive set-top box users were statistically insignificant in their 1999 report. William Mitchell, "Equitable Access to the Online World," in Schön, Sanyal, and Mitchell, eds., *High Technology and Low Income Community*, 151–62.

11. "Falling through the Net" (1995), 2–4.

12. "Falling through the Net II" (1998).

13. The E-Rate program has concerned itself only with students in public schools and largely focused on just getting machines into classrooms. But even this limited plan could not placate its opponents.

14. See Herbert Schiller, *Information Inequality: The Deepening Social Crisis in America* (New York: Routledge, 1996).

15. "Falling through the Net" (1999), 77–80.

16. James K. Glassman, "Gore's Internet Fiasco," *Washington Post*, June 2, 1998.

17. *Washington Post*, September 8, 1998.

18. "Issue Brief: Internet Access in Public Schools" (Washington, D.C.: National Center for Education Documents, 1998), 1. http://nces.ed.gov/pubs98/98031.html.

19. Tom Zeller, "Amid Clamor for Computer in Every Classroom, Some Dissenting Voices," *New York Times*, March 17, 1999, A18.

20. Boris Kachka, "Technology in Elementary Schools: Alternative Schools Encounter Computer Euphoria" (M.A. thesis, Columbia University Graduate School of Journalism, 1998).

21. Gregg B. Betheil, letter to the editor, *New York Times*, July 26, 1999, A14.

22. Sara Terry, "Across the Great Divide," *Fast Company*, July–August 1999, 192–200.

23. *Playing to Win—A Retrospective (1980–1994): Mission, History, Accomplishments* (1994), 1. Playing to Win is just one of the many impressive centers across the country.

24. For more information, see www.harlemlive.org and www.playingtowin.org.

25. Thomas P. Novak and Donna L. Hoffman, "Bridging the Digital Divide: The Impact of Race on Computer Access and Internet Use" (Vanderbilt University, February 2, 1998).

26. Ramon Harris, interview by author, January 19, 1999.

27. Keith Fulton, interview by author, January 21, 1999.

28. Peter Miller, "CTCNet and the Community Technology Center Movement" (1996), www.ctcnet.org.

29. Antonia Stone, "CTCNet: History, Organization, and Future," www.ctcnet.org.

30. Gary Chapman, "Reaching Out to Bring Low-Income Blacks across the 'Digital Divide,' " *Los Angeles Times*, April 12, 1999, C1.

31. Susan Myrland, "Bridging the Digital Divide in San Diego," *San Diego Union-Tribune*, July 13, 1999, B7, B9.

32. Benny Evangelista, "Helping Kids Go from Mean Streets to Information Superhighway," *San Francisco Chronicle*, July 5, 1999, C1. Antonia Stone has argued,

> The social and civic connections that participants make at community technology centers become increasingly important in a time when community ties are decreasing . . . and the locus of communication is . . . moving from the town square to the home. . . . It may be that community technology centers serve not only as places for accessing information and learning about technology but also as places to build and strengthen community ties, where people can make social connections and seek out others with similar interests.

Stone, "CTCNet: History, Organization, and Future."

33. James L. Tyson, "Latest Antipoverty Tool: A Computer," *Christian Science Monitor*, June 9, 1999, 4.

34. "Falling through the Net" (1999).

35. Blair Tindall, "Rainbow/PUSH Plans Bay Office: Jackson Group to Study 'Digital Divide,' " *San Francisco Examiner*, December 14, 1999, C1.

36. Ibid. The *San Jose Mercury News* found in April 1999 that there was only one black CEO leading a top 150 Silicon Alley firm, and just 8 percent of those companies had a person of Asian descent as chairman or CEO. Miguel Helft, "Symantec Names African American to CEO's Post," *San Jose Mercury News*, April 15, 1999.

37. Joshua Fishman, "The New Linguistic Order," *Foreign Policy*, December 22, 1998, 26.

38. Ted Bridis, "Minorities Fall through the Net: Report Warns of 'Racial Ravine' in Online Access," ABCNews.com, July 19, 1999.

39. Fernando Espuelas, interview by author, January 21, 1999.

40. Ibid.

41. Bill Gates, *The Road Ahead*, rev. ed. (New York: Penguin, 1996), 292.

42. Kirin Kalia, "Pod 8," *Silicon Alley Reporter*, online edition at www.siliconalleyreporter.com, May 1999.

43. Susan Goslee et al., *Losing Ground Bit by Bit: Low-Income Communities in the Information Age* (Washington, D.C.: Benton Foundation, 1998), iv, 2, 10.

44. The National Urban League and the Benton Foundation have found that the gaps in business can be particularly severe. Susan Goslee writes,

> Groups such as the United Church of Christ that have studied patterns

of telecommunications investment have found that, all too often, telephone and cable companies have moved quickly to wire wealthier suburbs with advanced systems, while poor, inner-city neighborhoods aren't upgraded. While public attention is often focused on whether individuals can get a service, the equally important problem is that lack of telecommunications facilities makes an area less attractive for businesses. This can feed a spiral where the lack of investment at the community level leads to fewer economic opportunities for people who live there. As a result, the poverty in the neighborhood makes it a less inviting target for investment, further aggravating the problem.

Goslee et al., *Losing Ground Bit by Bit*, iv.

45. Castells, *End of Millennium*, 74.

46. Harris, interview.

Chapter Two

"Their Logic against Them"

Contradictions in Sex, Race, and Class in Silicon Valley

Karen J. Hossfeld

> The bosses here have this type of reasoning like a seesaw. One day it's "you're paid less because women are different than men," or "immigrants need less to get by." The next day it's "you're all just workers here—no special treatment just because you're female or foreigners."
>
> Well, they think they're pretty clever with their doubletalk, and that we're just a bunch of dumb aliens. But it takes two to use a seesaw. What we're gradually figuring out here is *how to use their own logic against them.*
>
> —Filipina circuit board assembler in Silicon Valley (emphasis added)

This chapter examines how contradictory ideologies about sex, race, class, and nationality are used as forms of both labor control and labor resistance in the capitalist workplace today. Specifically, I look at the workplace relationships between Third World immigrant women production workers and their predominantly white male managers in high-tech manufacturing industry in Silicon Valley, California. My findings indicate that in workplaces where managers and workers are divided by sex and race, class struggle can and does take gender- and race-specific forms. Managers encourage women immigrant workers to identify with their gender, racial, and national identities when the managers want to "distract" the workers from their *class* concerns about working conditions. Similarly, when workers have workplace needs that actually *are* defined by gender, nationality, or race, managers tend to deny these identities and to stress the workers' generic class position. Immigrant women workers have learned to redeploy their managers' gender and racial tactics to their own advantage, however, in order to gain more control over their jobs. As the Filipina worker quoted at the beginning of the chapter so aptly said, they have learned to use managers' "own logic against them."

One of the objectives of this chapter is to expand traditional definitions of workplace resistance and control. All too frequently, these definitions have failed to consider the dynamics of gender and racial diversity. Another goal is to add to current theoretical debates about the changing conditions of capitalist, patriarchal, and national labor arrangements. My empirical data verify what Diane Elson and Ruth Pearson (1981). Maria Patricia Fernandez-Kelly (1983), June Nash and Fernandez-Kelly (1983), Helen Safa (1981), and many others have documented: namely, that the "new" international division of labor is increasingly based on gender, as well as class and nation. In addition, my findings confirm what Rachel Grossman (1979), Lenny Siegel (1980), and Linda Lim (1978) have each suggested: that high-tech industry is at the forefront of these trends toward a globalized, "gendered" labor division. Finally, I hope to inform strategic questions posed by organizers and workers who are faced with the struggle against this increasingly hierarchical and fragmented division of labor.

This chapter draws from a larger study of the articulation of sex, race, class, and nationality in the lives of immigrant women high-tech workers (Hossfeld 1988). Empirical data draw on more than two hundred interviews conducted between 1982 and 1986 with Silicon Valley workers; their family members, employers, and managers; and labor and community organizers. Extensive in-depth interviews were conducted with eighty-four immigrant women, representing twenty-one Third World nationalities, and with forty-one employers and managers, who represented twenty-three firms. All but five of these management representatives were U.S.-born white males. All of the workers and managers were employed in Santa Clara County, California, firms that engaged in some aspect of semiconductor "chip" manufacturing. I observed production at nineteen of these firms.

Before turning to the findings of my field research, I briefly situate the global context of Silicon Valley's highly stratified division of labor and identify the class structure and demographic features of its high-tech industry.

Silicon Valley

The Prototype

"Silicon Valley" refers to the microelectronics-based high-tech industrial region located just south of San Francisco in Santa Clara County, California. The area has been heralded as an economic panacea and as a regional prototype for localities around the globe that seek rapid economic growth and incorporation into the international market. Representatives from more than two thousand local and national governments, from People's Republic of China delegations to the queen of England, have visited the valley in search of a model for their own industrial revitalization. They have been awed by the sparkling, clean-looking facilities and the exuberant young executives who claim to have made riches overnight.[1] But the much-fetishized Silicon Valley "model" that so many seek to emulate implies more than just the potential promise of jobs, revenue, growth, and participation in the technological "revolution."

The development of the microelectronics industry in its current form involves highly problematic relations of production that extend both inside and outside the workplace. Microelectronics is the "way of the future" not only technologically but, as developed under capitalism, in its work arrangements and social relationships, which are predicated on sharp divisions according to sex, race, class, and nation. Not only the *technology* of microelectronics but the structure of its industries as well are important tools in the capitalist economy's constant search for new permutations in the division of labor. What Silicon Valley is all about, then, is more than laser technology, video games, and illusory hot tubs for the masses; it is about class structure, class struggle, and the division of labor.

The media have been much enamored with the computer "revolution" in general and in particular with the imagery and ideology of the industry's preponderance of "self-made" millionaires. But for every young, white boy-wonder who made his first million tinkering in the garage (as folklore says the founders of the Apple Computer Company, Steve Jobs and Steve Wozniak, did), there are scores of low-paid immigrant women workers. These women from Mexico, China, Vietnam, Korea, the Philippines, and other Third World countries prop up the computer revolution, in what amounts to a very *un*revolutionary industrial division of labor.

Since the 1960s, the large U.S. microelectronics manufacturers have been shifting production facilities to "offshore" locations, primarily in Southeast Asia, but also in Mexico, Puerto Rico, and other locations in both the Third World and Europe. Assembly work has been particularly easy to shift abroad since the materials involved are light in weight, small in size, and of relatively low per-unit value. Materials and assembled products are thus easy to transport. Assembly requires little special equipment or skilled labor when performed manually: the work involves relatively low capital investment but is labor-intensive. Frequently, semiconductors are manufactured in the United States, shipped abroad, where they are assembled and sometimes tested, and then shipped back to the United States for final inspection, packaging, and distribution. Increasingly, any one circuit board on the U.S. market represents labor performed in several countries.

The major motivating force in shifting production has been to cut labor costs. Assembly workers in most Third World high-tech outposts are paid only a few dollars a day. According to my informants at the American Electronics Association, assembly workers in the Philippines, for example, earn one-tenth of what they do in Silicon Valley. But which countries the companies choose to go to has depended on other labor and investment considerations as well. The Singapore government, for example, has actively courted multinational high-tech firms and has offered them economic and tax incentives. The governments of Singapore, Korea, and the Philippines have strongly discouraged and in some cases outlawed labor organizing and strikes. In a global economy that is short on both capital and jobs, microelectronics firms have been able to shop around the world for the most advantageous labor market conditions.

The skewed division of labor based on gender and race is even more pronounced in offshore sites than in Silicon Valley. Invariably, offshore assembly plants employ young women almost exclusively, the figurative if not the literal sisters of Third World immigrants who work in Silicon Valley. With no Equal Employment Opportunity Commission (EEOC) or other state watchdogs and no unions with bargaining power, firms are free to discriminate openly according to sex, age, and marital status.

Although most low-paying, high-tech jobs are sent to the periphery and semiperiphery of the world economy, firms continue to employ production workers in Silicon Valley to meet the ongoing need for quickly available prototypic and short-term projects. Thus, during the 1980s, the *percentage* of production work done in the valley declined drastically, as the industry expanded, but the number of production jobs did not decrease significantly. Although most manufacturing done on-site in Silicon Valley involves "higher-tech" stages of the production process, the assembly work that immigrant women engage in closely resembles the same "low-tech" labor done by their "sisters" overseas.

What Silicon Valley offers the numerous other "Silicon dales" and "deserts" that are springing up around the world is not only a model of technological know-how and development, but also—and equally important—a model of labor control based on highly stratified divisions of race, sex, and class. This study suggests that this division of labor stratifies not only Singapore and San Francisco—core and periphery—but workers within the core metropole itself.

Class Structure and the Division of Labor

Close to 200,000 people—one out of every four employees in the San Jose Metropolitan Statistical Area labor force—work in Silicon Valley's microelectronics industry. There are more than 800 manufacturing firms that hire ten or more people each, including 120 "large" firms that each count over 250 employees. An even larger number of small firms hire fewer than ten employees apiece. Approximately half of this high-tech labor force—100,000 employees—works in production-related work: at least half of these workers—an estimated 50,000 to 70,000—are in low-paying, semiskilled operative jobs (Siegel and Borock 1982; *Annual Planning Information* 1983).[2]

The division of labor within the industry is dramatically skewed according to gender and race. Although women account for close to half of the total paid labor force in Santa Clara County both inside and outside the industry, only 18 percent of the managers, 17 percent of the professional employees, and 25 percent of the technicians are female. Conversely, women hold at least 68 percent and by some reports as many as 85 to 90 percent of the valley's high-tech operative jobs. In the companies examined in my study, women made up an average of 90 percent of the assembly and operative workers. Only rarely do they work as production managers or supervisors, the management area that works most closely with the operatives.

Similar disparities exist vis-à-vis minority employment. According to the 1980 census, 26.51 percent of the civilian workforce of Santa Clara County was composed of racial minorities. Fifteen percent were Hispanic (all races); 7.5 percent were Asian–Pacific Islanders; 3 percent were Black; 0.5 percent were Native American; and 0.2 percent were listed as "other races—not Hispanic" (*Annual Planning Information* 1983, 96–97). Over 75 percent of the Hispanics were of Mexican descent. Of the 102,000 Asian–Pacific Islanders counted in the 1980 census as living in the area, roughly 28 percent were Filipino or of Filipino descent; 22 percent each were Japanese and Chinese; 11 percent were Vietnamese; 6 percent were Korean; 5 percent were Asian Indian; and less than 2 percent each were of other national origins (*Annual Planning Information* 1983, 64).

Since the census was taken, influxes of refugees from Indochina have quadrupled the number of Vietnamese, Laotians, and Cambodians in the area: as of early 1984, there were an estimated forty-five thousand Southeast Asian refugees in Santa Clara County, as well as a smaller but growing number of refugees from other regions such as Central America. I have talked with Silicon Valley production workers from at least thirty Third World nations. In addition to the largest groups, whose members are from Mexico, Vietnam, the Philippines, and Korea, workers hail from China, Cambodia, Laos, Thailand, Malaysia, Indonesia, India, Pakistan, Iran, Ethiopia, Haiti, Cuba, El Salvador, Nicaragua, Guatemala, and Venezuela. There are also small groups from southern Europe, particularly Portugal and Greece.

Within the microelectronics industry, 12 percent of the managers,

16 percent of the professionals, and 18 percent of the technicians are minorities—although they are concentrated at the lower-paying and less powerful ends of these categories. An estimated 50 to 75 percent of the operative jobs are thought to be held by minorities.[3] My study suggests that the figure may be closer to 80 percent.

Both employers and workers interviewed in this study agreed that the lower the skill and pay level of the job, the higher the percentage of Third World immigrant women who were employed. Thus assembly work, which is the least skilled and lowest-paying production job, tends to be done predominantly by Third World women. Entry level production workers, who work in job categories such as semiconductor processing and assembly, earn an average of $4.50 to 5.50 an hour; experienced workers in these jobs earn from $5.50 to $8.50. At the subcontracting assembly plants I observed, immigrant women accounted for 75 to 100 percent of the production labor force. At only one of these plants did white males account for more than 2 percent of the production workers. More than 90 percent of the managers and owners at these businesses were white males, however.

This occupational structure is typical of the industry's division of labor nationwide. The percentage of women of color in operative jobs is fairly standardized throughout various high-tech centers; what varies is *which* minority groups are employed, not the job categories in which they are employed.[4]

Obviously, there is tremendous cultural and historical variation both between and within the diverse national groups that my informants represent. Here I emphasize their commonalities. Their collective experience is based on their jobs, present class status, recent uprooting, and immigration. Many are racial and ethnic minorities for the first time. Finally, they have in common their gender and their membership in family households.

Labor Control on the Shop Floor

Gender and Racial Logic

In Silicon Valley production shops, the ideological battleground is an important arena of class struggle for labor control. Management frequently calls upon ideologies and arrangements concerning sex and race, as well as class, to manipulate worker consciousness and to legitimate the hierarchical division of labor. Management taps both traditional popular stereotypes about the presumed lack of status and limited abilities of women, minorities, and immigrants and the workers' own fears, concerns, and sense of priorities as immigrant women.

But despite management's success in disempowering and devaluing labor, immigrant women workers have co-opted some of these ideologies and have developed others of their own, playing on management's prejudices to the workers' own advantage. In so doing, the workers turn the "logic" of capital against managers, as they do the intertwining logics of patriarchy and racism. The following section examines this sex- and race-based logic and how it affects class structure and struggle. I then focus on women's resistance to this manipulation and their use of gender and racial logics for their own advantage.

From interviews with Silicon Valley managers and employers, it is evident that high-tech firms find immigrant women particularly appealing workers not only because they are "cheap" and considered easily "expendable" but also because management can draw on and further exploit preexisting patriarchal and racist ideologies and arrangements that have affected these women's consciousness and realities. In their dealings with the women, managers fragment the women's multifaceted identities into falsely separated categories of "worker," "ethnic," and "woman." The effect is to increase and play off the workers' vulnerabilities and splinter their consciousness. But I also found limited examples of the women drawing strength from their multifaceted experiences and developing a unified consciousness with which to confront their oppressions. These instances of how the workers have manipulated management's ideology are important not only in their own right but as models. To date, though, management holds the balance of power in this ideological struggle.

I label management's tactics "gender-specific" and "racial-specific" forms of labor control and struggle, or gender and racial "logic." I use the term *capital logic* to refer to strategies by capitalists to increase profit maximization. Enforcement by employers of a highly stratified class division of labor as a form of labor control is one such strategy. Similarly, I use the terms *gender logic* and *racial logic* to refer to strategies to promote gender and racial hierarchies. Here I am concerned primarily with the ways employers and managers devise and incorporate gender and racial logic in the interests of capital logic. Attempts to legitimate inequality form my main examples.

I focus primarily on managers' "gender-specific" tactics because management uses race-specific (il)logic much less directly in dealing with workers. Management clearly draws on racist assumptions in hiring and dealing with its workforce, but usually it makes an effort to conceal its racism from workers. Management recognizes, to varying degrees, that the appearance of blatant racism against workers is not acceptable, mainly because immigrants have not sufficiently internalized racism to respond to it positively. Off the shop floor, however, the managers' brutal and open racism toward workers was apparent during "private" interviews. Managers' comments demonstrate that racism is a leading factor in capital logic but that management typically disguises racist logic by using the more socially acceptable "immigrant logic." Both American and immigrant workers tend to accept capital's relegation of immigrants to secondary status in the labor market.

Conversely, "gender logic" is much less disguised: management uses it freely and directly to control workers. Patriarchal and sexist ideology is *not* considered inappropriate. Because women workers themselves have already internalized patriarchal ideology, they are more likely to "agree" with or at least accept it than they are racist assumptions. This chapter documents a wide range of sexist assumptions that management employs in order to control and divide workers.

Gender Ideology

A growing number of historical and contemporary studies illustrate the interconnections between patriarchy and capitalism in defining both the daily lives of working women and the nature of work arrangements in general. Sallie Westwood, for example, suggests that on-the-job exploitation of women workers is rooted in part in patriarchal ideology. Westwood states that ideologies "play a vital part in calling forth a sense of self linked to class and gender as well as race. Thus, a patriarchal ideology intervenes on the shopfloor culture to make anew the conditions of work under capitalism" (1985–6).

One way patriarchal ideology affects workplace culture is through the "gendering" of workers—what Westwood refers to as "the social construction of masculinity and femininity on the shop floor" (6). The forms of work culture that managers encourage, and that women workers choose to develop, are those that reaffirm traditional forms of femininity. This occurs in spite of the fact that, or more likely because, the women are engaged in roles that are traditionally defined as nonfeminine: factory work and wage earning. My data suggest that although factory work and wage earning are indeed traditions long held by working-class women, the dominant *ideology* that such tasks are "unfeminine" is equally traditional. For example, I asked one Silicon Valley assembler who worked a double shift to support a large family how she found time and finances to obtain elaborate manicures, makeup, and hair stylings. She said that they were priorities because they "restored [her] sense of femininity." Another production worker said that factory work "makes me feel like I'm not a lady, so I have to try to compensate."

This ideology about what constitutes proper identity and behavior for women is multileveled. First, women workers have a clear sense that wage earning and factory work in general are not considered "feminine." This definition of "feminine" derives from an upperclass reality in which women traditionally did not need (and men often did not allow them) to earn incomes. The reality for a production worker who comes from a long line of factory women does not negate the dominant ideology that influences her to say, "At work I feel stripped of my womanhood. I feel like I'm not a lady anymore. It makes me feel . . . unattractive and unfeminine."

Second, women may feel "unwomanly" at work because they are away from home and family, which conflicts with ideologies, albeit changing ones, that they should be home. And third, earning wages at all is considered "unwifely" by some women, by their husbands, or both because it strips men of their identity as "breadwinner."

On the shop floor, managers encourage workers to associate "femininity" with something contradictory to factory work. They also encourage women workers to "compensate" for their perceived loss of femininity. This strategy on the part of management serves to devalue women's productive worth.

Under contemporary U.S. capitalism, ideological legitimation of women's societal roles and of their related secondary position in the division of labor is already strong outside the workplace. Management thus does not need to devote extreme efforts to developing *new* sexist ideologies within the workplace in order to legitimate the gender division of labor. Instead, managers can call on and reinforce preexisting ideology. Nonetheless, new forms of gender ideology are frequently introduced. These old and new ideologies are disseminated both on an individual basis, from a manager to a worker or workers, and on a collective basis, through company programs, policies, and practices. Specific examples of informal ways individual managers encourage gender identification, such as flirting, dating, sexual harassment, and promoting "feminine" behavior, are given below. The most widespread company practice that encourages engenderment, of course, is hiring discrimination and job segregation based on sex.

An example of a company policy that divides workers by gender is found in a regulation one large firm has regarding color-coding of smocks that all employees in the manufacturing division are required to wear. While the men's smocks are color-coded according to occupation, the women's are color-coded by sex, regardless of occupation. This is a classic demonstration of management's encouragement of male workers to identify according to job and class and its discouragement of women from doing the same. Regardless of what women do as workers, the underlying message reads, they are nevertheless primarily women. The same company has other practices and programs that convey the same message. The company newsletter, for example, includes a column entitled "Ladies' Corner," which runs features on cooking and fashion tips for "the working gal." A manager at this plant says that such "gender tactics," as I call them, are designed to "boost morale by reminding the gals that even though they do unfeminine work, they really are still feminine." But although some women workers may value femininity, in the work world, management identifies feminine traits as legitimation for devaluation.

In some places, management offers "refeminization" perks to help women feel "compensated" for their perceived "defeminization" on the job. A prime example is the now well-documented makeup sessions and beauty pageants for young women workers sponsored by multinational electronics corporations at their Southeast Asian plants (Grossman 1979; Ong 1985). While such events are unusual in Silicon Valley, male managers frequently use flirting and dating as "refeminization" strategies. Flirting and dating in and of themselves certainly cannot be construed as capitalist plots to control workers; however, when they are used as false compensation for and to divert women from poor working conditions and workplace alienation, they in effect serve as a form of labor control. In a society where women are taught that their femininity is more important than other aspects of their lives—such as how they relate to their work—flirting can be divisive. And when undesired, flirting can also develop into a form of sexual harassment, which causes further workplace alienation.

One young Chinese production worker told me that she and a coworker avoided filing complaints about illegal and unsafe working conditions because they did not want to annoy their white male supervisor, whom they enjoyed having flirt with them. These two women would never join a union, they told me, because the same supervisor told them that all women who join unions "are a bunch of tough, big-mouthed dykes." Certainly these women have the option of ignoring this man's opinions. But that is not easy, given the one-sided power he has over them not only because he is their supervisor, but because of his age, race, and class.

When women workers stress their "feminine" and female characteristics as being counter to their waged work, a contradictory set of results can occur. On one hand, the women may legitimate their own devaluation as workers, and, in seeking identity and solace in their "femininity," discard any interest in improving their working conditions. On the other hand, if turning to their identities as female, mother, mate, and such allows them to feel self-esteem in one arena of their lives, that self-esteem may transfer to other arenas. The outcome is contingent on the ways the women define and experience themselves as female or "feminine." Femininity in white American capitalist culture is traditionally defined as passive, and ineffectual, as Susan Brownmiller explores (1984). But there is also a female tradition of resistance.

The women I interviewed rarely pose their womanhood or their self-perceived femininity as attributes meriting higher pay or better treatment. They expect *differential* treatment because they are women, but "differential" inevitably means lower paid in the work world. The women present their self-defined female attributes as creating additional needs that detract from their financial value. Femininity, although its definition varies among individuals and ethnic groups, is generally viewed as something that subtracts from a woman's market value, even though a majority of women consider it personally desirable.

In general, both the women and men I interviewed believe that women have many needs and skills discernible from those of male workers, but they accept the ideology that such specialness renders them less deserving than men of special treatment, wages, promotions, and status. Conversely, both the men and women viewed men's special needs and skills as rendering men *more* deserving. Two of the classic perceived sex differentials cited by employers in electronics illustrate this point. First, although Silicon Valley employers consistently repeat the old refrain that women are better able than men to perform work requiring manual skills, strong hand-eye coordination, and extreme patience, they nonetheless find it appropriate to pay workers who have these skills (women) less than workers who supposedly do not have them (men). Second, employers say that higher entry-level jobs, wages, and promotions rightly belong to heads of households, but in practice they give such jobs only to men, regardless of their household situation, and exclude women, regardless of theirs.

When a man expresses special needs that result from his structural position in the family—such as head of household—he is often "compensated," yet when a woman expresses a special need resulting from her traditional structural position in the family—child care or *her* position as head of household—she is told that such issues are not of concern to the employer or, in the case of child care, that it detracts from her focus on her work and thus devalues her productive contribution. This is a clear illustration of Heidi Hartmann's definition of patriarchy: social relationships between men, which, although hierarchical, such as those between employer and worker, have a material base that benefits men and oppresses women (1976).

Definitions of femininity and masculinity not only affect the workplace but are in turn affected by it. Gender is produced and reproduced in and through the workplace, as well as outside it. Gender identities and relationships are formed on the work floor both by the labor process organized under capitalism and by workers' resistance to that labor process. "Femininity" in its various permutations is not something all-bad and all-disempowering: women find strength, pride, and creativity in some of its forms. But the ideological counterpositioning of "feminine" as weak, powerless, and submissive and of "masculine" as strong, powerful, and dominant raises yet another problem in the workplace: sexual harassment. For reasons of space. I will not discuss this here. I turn now to one of the other tenets of women workers' multitiered consciousness that employers find advantageous: gender logic that poses women's work as "secondary."

The Logic of "Secondary" Work

Central to gender-specific capital logic is the assumption that women's paid work is both secondary and temporary. More than 70 percent of the employers and 80 percent of the women workers I interviewed stated that a woman's primary jobs are those of wife, mother, and homemaker, even when she works full-time in the paid labor force. Because employers view women's primary job as in the home, and they assume that, prototypically, every woman is connected to a man who is bringing in a larger paycheck, they claim that women do not need to earn a full living wage. Employers repeatedly asserted that they believed the low-level jobs were filled only by women because men could not afford to or would not work for such low wages.

Indeed, many of the women would not survive on what they earned unless they pooled resources. For some, especially the nonimmigrants, low wages did mean dependency on men—or at least on family networks and household units. None of the women I interviewed—immigrant or nonimmigrant—lived alone. Yet most of them would be financially better off without their menfolk. For most of the immigrant women, their low wages were the most substantial and steady source of their family's income. *Eighty percent of the immigrant women workers in my study were the largest per annum earners in their households.*

Even when their wages were primary—the main or only family income—the women still considered men to be the major breadwinners. The women considered their waged work as secondary, both in economic value and as a source of identity. Although most agreed that women and men who do exactly the same jobs should be paid the same, they had little expectation that as women they would be eligible for higher-paying "male" jobs. While some of these women—particularly the Asians—believed they could overcome racial and class barriers in the capitalist division of labor, few viewed gender as a division that could be changed. While they may believe that hard work can overcome many obstacles and raise their *families'* socioeconomic class standing, they do not feel that their position in the gender division of labor will change. Many, of course, expect or hope for better jobs for themselves—and others expect or hope to leave the paid labor force altogether—but few wish to enter traditional male jobs or to have jobs that are higher in status or earnings than the men in their families.

The majority of women who are earning more than their male family members view their situation negatively and hope it will change soon. They do not want to earn less than they currently do; rather, they want their menfolk to earn more. This was true of women in all the ethnic groups. The exceptions—a vocal minority—were mainly Mexicans. Lupe, a high-tech worker in her twenties, explained:

> Some of the girls I work with are ridiculous—they think if they earn more than their husbands it will hurt the men's pride. They play up to the machismo. . . . I guess it's not entirely ridiculous, because some of them regularly come in with black eyes and bruises, so the men are something they have to reckon with. But, my God, if I had a man like that I would leave. . . .
>
> My boyfriend's smart enough to realize that we need my paycheck to feed us and my kids. He usually brings home less than I do, and we're both damn grateful for every cent that either of us makes. When I got a raise he was very happy—I think he feels more relieved, not more resentful. But then, he's not a very typical man, no? Anyway, he'd probably change if we got married and had kids of his own—that's when they start wanting to be the king of their castle.

A Korean immigrant woman in her thirties told how her husband was so adamant that she not earn more than he and that the men in the household be the family's main supporters that each time she cashed her paycheck she gave some of her earnings to her teenaged son to turn over to the father as part of his earnings from his part-time job. She was upset about putting her son in a position of being deceitful to his father, but both mother and son agreed it was the only alternative to the father's otherwise dangerous, violent outbursts.

As in the rest of America, in most cases, the men earned more in those households where both the women and men worked regularly. In many of the families, however, the men tended to work less regularly than the women and to have higher unemployment rates. While most of the families vocally blamed very real socioeconomic conditions for the unemployment, such as declines in "male" industrial sector jobs, many women also felt that their husbands took out their resentment on their families. A young Mexicana, who went to a shelter for battered women after her husband repeatedly beat her, described her extreme situation:

> He knows it's not his fault or my fault that he lost his job: they laid off almost his whole shift. But he acts like I keep my job just to spite him, and it's gotten so I'm so scared of him. Sometimes I think he'd rather kill me or have us starve than watch me go to work and bring home pay. He doesn't want to hurt me, but he is so hurt inside because he feels he has failed as a man.

Certainly not all laid-off married men go to the extreme of beating their wives, but the majority of married women workers whose husbands had gone through periods of unemployment said that the men treated other family members significantly worse when they were out of work. When capitalism rejects male workers, they often use patriarchal channels to vent their anxieties. In a world where men are defined by their control over their environment, losing control in one arena, such as that of the work world, may lead them to tighten control in another arena in which they still have power—the family. This classic cycle is not unique to Third World immigrant communities, but as male unemployment increases in these communities, so may the cycle of male violence.

Even some of the women who recognize the importance of their economic role feel that their status and identity as wage earners are less important than those of men. Many of the women feel that men work not only for income but for respect and dignity. They see their own work as less noble. Although some said they derive satisfaction from their ability to hold a job, none of the women considered her job to be a primary part of her identity or a source of self-esteem. These women see themselves as responsible primarily for the welfare of their families: their main identity is as mother, wife, sister, and daughter, not as worker. Their waged work is seen as an extension of caring for their families. It is not a question of *choosing* to work—they do so out of economic necessity.

When I asked whether their husbands' and fathers' waged work could also be viewed as an extension of familial duties, the women indicated that they definitely perceived a difference. Men's paid labor outside the home was seen as integral both to the men's self-definition and to their responsibility vis-à-vis the family; conversely, women's labor force participation was seen as contradictory both to the women's self-image and to their definitions of female responsibility.

Many immigrant women see their wage contribution to the family's economic survival not only as secondary but as *temporary*, even when they have held their jobs for several years. They expect to quit their production jobs after they have saved enough money to go to school, stay home full-time, or open a family business. In actually, however, most of them barely earn enough to live on, let alone to save, and women who think they are signing on for a brief stint may end up staying in the industry for years.

That these workers view their jobs as temporary has important ramifications for both employers and unions, as well as for the workers themselves. When workers believe they are on board a company for a short time, they are more likely to put up with poor working conditions, because they see them as short-term. A Mexican woman who used to work in wafer fabrication reflected on the consequences of such rationalization:

> I worked in that place for four years, and it was really bad—the chemicals knocked you out, and the pay was very low. My friends and me, though, we never made a big deal about it, because we kept thinking we were going to quit soon anyway, so why bother . . . We didn't really think of it as our career or anything—just as something we had to do until our fortune changed. It's not exactly the kind of work a girl dreams of herself doing.
>
> My friend was engaged when we started working there, and she thought she was going to get married any day, and then she'd quit. Then, after she was married, she thought she'd quit as soon as she got pregnant . . . She has two kids now, and she's still there. Now she's saying she'll quit real soon, because her husband's going to get a better job any time now, and she'll finally get to stay home, like she wants.

Ironically, these women's jobs may turn out to be only temporary, but for different reasons and with different consequences than they planned. Industry analysts predict that within the next decade the majority of Silicon Valley production jobs may well be automated out of existence (Carey 1984). Certainly for some of the immigrant women, their dreams of setting aside money for occupational training or children's schooling or to open a family business or finance relatives' immigration expenses do come true, but not for most. Nonetheless, almost without exception, the women production workers I interviewed—both immigrant and nonimmigrant—saw their present jobs as temporary.

Employers are thus at an advantage in hiring these women at low wages and with little job security. They can play on the women's *own* consciousness as wives and mothers whose primary identities are defined by home and familial roles. While the division of labor prompts the workers to believe that women's waged work is less valuable than men's, the women workers themselves arrive in Silicon Valley with this ideology already internalized.

A young Filipina woman, who was hired at a walk-in interview at an electronics production facility, experienced a striking example of the contradictions confronting immigrant women workers in the valley. Neither she nor her husband, who was hired the same day, had any previous related work experience or degrees. Yet her husband was offered an entry-level job as a technician, while she was offered an assembly job paying three dollars per hour less. The personnel manager told her husband that he would "find [the technician job] more interesting than assembly work." The woman had said in the interview that she wanted to be considered for a higher-paying job because she had two children to support. The manager refused to consider her for a different job, she said, and told her that "it will work out fine for you, though, because with your husband's job, and you *helping out* [emphasis added] you'll have a nice little family income."

The same manager told me on a separate occasion that the company preferred to hire members of the same families because it meant that workers' relatives would be more supportive about their working and the combined incomes would put less financial strain on individual workers. This concern over workers and their families dissipated, however, when the Filipino couple split up, leaving the wife with only the "helping-out" pay instead of the "nice little family income." When the woman requested a higher-paying job so she could support her family, the same manager told her that "family concerns were out of place at work" and did not promote her.

This incident suggests that a woman's family identity is considered important when it is advantageous to employers and irrelevant when it is disadvantageous. Similarly, managers encourage women workers to identify themselves primarily as workers or as women, depending on the circumstances. At one plant where I interviewed both managers and workers, males and females were openly separated by the company's hiring policy: entry-level jobs for females were in assembly, and entry-level jobs for males were as technicians. As at the plant where the Filipino couple worked, neither the "male" nor the "female" entry-level jobs required previous experience or training, but the "male" job paid significantly more.

Apparently, the employers at this plant *did* see differences between male and female workers, despite their claims to the contrary. Yet, when the women workers asked for "special treatment" because of these differences, the employers' attitudes rapidly changed. When the first quality circle was introduced in one production unit at this plant, the workers, all of whom were women, were told to suggest ways to improve the quality of work.[5] The most frequently mentioned concern of all the women production workers I met was the lack of decent child care facilities. The company replied that child care was not a quality of work–related issue but a "special women's concern" that was none of the company's business.

A Portuguese worker succinctly described the tendency among employers to play on and then deny such gender logic:

> The boss tells us not to bring our "women's problems" with us to work if we want to be treated equal. What does he mean by that? I am working here *because* of my "women's problems"—because I am a woman. Working here *creates* my "women's problems." I need this job because I am a woman and have children to feed. And I'll probably get fired because I am a woman and need to spend more time with my children. I am only one person—and I bring my whole self to work with me. So what does he mean, don't bring my "women's problems" here?

As this woman's words so vividly illustrate, divisions of labor and of lives are intricately interwoven. Any attempts to organize the women workers of Silicon Valley—by unions, communities, political or social groups and by the women themselves—must deal with the articulation of gender, race, and class inequalities in their lives.

Resistance on the Shop Floor

There is little incidence in Silicon Valley production shops of *formal* labor militancy among the immigrant women, as evidenced by either union participation or collectively planned mass actions such as strikes. Filing formal grievances is not common in these workers' shop culture. Union activity is very limited, and both workers and managers claim that the incidence of complaints and disturbances on the shop floor is lower than in other industries. Pacing of production to restrict output does occur, and there are occasionally "informal" incidents, such as spontaneous slowdowns and sabotage. But these actions are rare and usually small in scale. Definitions of workplace militancy and resistance vary, of course, according to the observer's cultural background, but by their *own* definitions, the women do not frequently engage in traditional forms of labor militancy.

There is, however, an important, although often subtle, arena in which the women do engage in struggle with management: the ideological battleground. Just as employers and managers harness racist, sexist, and class-based logic to manipulate and control workers, so too workers use this logic against management. In the ideological arena, the women do not merely accept or react to the biased assumptions of managers: they also develop gender-, class-, and race-based logic of their own when it is to their advantage. The goal of these struggles is not simply ideological victory but concrete changes in working conditions. Further, in Silicon Valley, immigrant women workers have found that managers respond more to workers' needs when they are couched in ethnic or gender terms, rather than in class and labor terms. Thus, class struggle on the shop floor is often disguised as arguments about the proper place and appropriate behavior of women, racial minorities, and immigrants.

When asked directly, immigrant women workers typically deny that they engage in any form of workplace resistance or efforts to control their working conditions. This denial reflects not only workers' needs to protect clandestine activities, but also their consciousness about what constitutes resistance and control. In their conversations with friends and coworkers, the women joke about how they outfoxed their managers with female or ethnic "wisdom." Yet most of the women do not view their often elaborate efforts to manipulate their managers' behavior as forms of struggle. Rather, they think of their tactics "just as ways to get by," as several workers phrased it. It is from casual references to these tactics that a portrait of worker logic and resistance emerges.

The workers overwhelmingly agreed that the challenges to management in which they could and did engage were on a small-scale, individual, or small-group level. Several women said they engaged in forms of resistance that they considered "quiet" and unobtrusive: acts that would make a difference to the woman and possibly her coworkers but that management would probably not recognize as resistance. Only rarely was resistance collectively articulated.

The vast majority of these women clearly wish to avoid antagonizing management. Thus, rather than engaging in confrontational resistance strategies, they develop less obvious forms than, say, work stoppages, filing grievances, and straightforwardly refusing to perform certain tasks, all of which have frequently been observed in other industrial manufacturing sectors. Because the more "quiet" forms of resistance and struggle for workplace control engaged in by the women in Silicon Valley are often so discrete and the workers are uncomfortable discussing them, it is probable that there are more such acts and they are broader in scope than my examples imply. As a Chinese woman in her forties who has worked as an operative in the valley for six years explained,

> Everybody who does this job does things to get through the day, to make it bearable. There are some women who will tell you they never do anything unproper or sneaky, but you are not to believe them. The ones that look the most demure are always up to something. . . . There's not anybody here who has never purposefully broken something, slowed down work, told fibs to the supervisor, or some such thing. And there's probably no one but me with my big mouth who would admit it!

As discussed above, it is clear that managers have found effective ways to play off workers' gender, racial, and immigrant consciousness. At the same time, white male managers in particular often have striking misconceptions about the gender and cultural experiences of their workers, and workers can thus frequently confuse them with bogus claims about the women's special needs. Workers can also use real claims that supervisors have tried to co-opt.

A Salvadorean woman, fed up with her supervisor for referring to his Hispanic workers as "mamacitas" and "little mothers" and admonishing them to "work faster if you want your children to eat," had her husband bring both her own children and several nieces and nephews to pick her up one day. She lined all the children up in front of the supervisor and asked him how fast she would have to work to feed all those mouths. One of the children had been coached, and he told the supervisor that his mother was so tired from working that she did not have time to play with them anymore. The guilt-ridden supervisor, astonished by the large number of children and the responsibility they entailed, eased up on his admonishments and speed-up efforts and started treating the woman with more consideration.

The most frequently mentioned acts of resistance against management and work arrangements were ones that played on the white male managers' consciousness—both false and real—about gender and ethnic culture. Frequently mentioned examples involved workers who turned management's ideologies against them by exploiting their male supervisors' misconceptions about "female problems." A white chip tester testified:

> It's pretty ironic because management seems to have this idea that male supervisors handle female workers better than female supervisors. You know, we're supposed to turn to mush whenever he's around and respect his authority or something. But this one guy we got now lets us walk all over him. He thinks females are flighty and irresponsible because of our hormones—so we make sure to have as many hormone problems as we can. I'd say we each take hormone breaks several times a day. My next plan is to convince him that menstrual blood will turn the solvents bad, so on those days we have to stay in the lunchroom!

A Filipina woman production worker recounted another example:

> The boss told us girls that we're not strong enough to do the heavy work in the men's jobs—and those jobs pay more, too. So, I suddenly realized that gosh, us little weak little things shouldn't be lifting all those heavy boxes of circuit board parts we're supposed to carry back and forth all the time—and I stopped doing it.
>
> The boss no longer uses that "it's too heavy for you girls" line anymore . . . but I can tell he's working on a new one. That's okay: I got plenty of responses.

A Mexican wafer fabricator, whose unit supervisor was notorious for the "refeminization" perks discussed above, told of how she manipulated the male supervisor's gender logic to disguise what was really an issue of class struggle:

> I was getting really sick from all the chemicals we have to work with, and I was getting a rash from them on my arms. [The manager] kept saying I was exaggerating and gave the usual line about you can't prove what caused the rash. One day we had to use an especially harsh solvent, and I made up this story about being in my sister's wedding. I told him that the solvents would ruin my manicure, and I'd be a mess for the wedding. Can you believe it? He let me off the work! This guy wouldn't pay attention to my rash, but when my manicure was at stake, he let me go!

Of course, letting this worker avoid chemicals for one day because of a special circumstance is more advantageous to management than allowing her and others to avoid the work permanently because of health risks. Nonetheless, the worker was able to carve out a small piece of bargaining power by playing off her manager's gender logic. The contradiction of these tactics that play up feminine frailty is that they achieve short-term, individual goals at the risk of reinforcing damaging stereotypes about women, including the stereotype that women workers are not as productive as men. From the workers' point of view, however, the women are simply using the prejudices of the powerful to the advantages of the weak.

Another "manicure" story resulted in a more major workplace change at one of the large plants. Two women fabricator operatives, one Portuguese and one Chicana, applied for higher-paying technician jobs whereupon their unit supervisor told them that the jobs were too "rough" for women and that the work would "ruin their nails." The women's response was to pull off their rubber gloves and show him what the solvents and dopants had done to their nails, despite the gloves. (One of the most common chemicals used in chip manufacturing is acetone, the key ingredient in nail polish removal. It also eats right through "protective" rubber gloves.) After additional goading and bargaining, the supervisor provisionally let them transfer to technician work.

Although the above are isolated examples, they represent tactics that workers can use either to challenge or play off sexist ideology that employers use to legitimate women's low position in the segregated division of labor. Certainly there are not enough instances of such behavior to challenge the inequality between worker and boss, but they do demonstrate to managers that gender logic cannot always be counted on to legitimate inequality between male and female workers. And dissolving divisions between workers *is* a threat to management hegemony.

Racial and Ethnic Logic

Typically, high-tech firms in Silicon Valley hire production workers from a wide spectrum of national groups. If their lack of a common language (both linguistically and culturally) serves to fragment the labor force, capital benefits. Conversely, management may find it more difficult to control workers with whom it cannot communicate precisely. Several workers said they have feigned a language barrier in order to avoid taking instructions; they have also called forth cultural taboos—both real and feigned—to avoid undesirable situations. One Haitian woman, who took a lot of kidding from her employer about voodoo and black magic, insisted that she could not work the night shift because evil spirits were out then. Because she was a good worker, the employer let her switch to days. When I tried to establish whether she believed the evil spirits were real or imagined, she laughed and said, "Does it matter? The result is the same: I can be home at night with my kids."

Management in several plants believed that racial and national diversity minimized solidarity. According to one supervisor, workers were forbidden from sitting next to people of their own nationality (i.e., language group) in order to "cut down on the chatting." Workers quickly found two ways to reverse this decision, using management's own class, racial, and gender logic. Chinese women workers told the supervisor that if they were not "chaperoned" by other Chinese women, their families would not let them continue to work there. Vietnamese women told him that the younger Vietnamese women would not work hard unless they were under the eyes of the older workers and that a group of newly hired Vietnamese workers would not learn to do the job right unless they had someone who spoke their language to explain it to them. Both of these arguments could also be interpreted as examples of older workers wanting to control younger ones in a generational hierarchy, but this was not the case. Afterwards both the Chinese and the Vietnamese women laughed among themselves at their cleverness. Nor did they forget the support needs of workers from other ethnic groups: they argued with the supervisor that the same customs and needs held true for many of the language groups represented, and the restriction was rescinded.

Another example of a large-scale demonstration of interethnic solidarity on the shop floor involved workers playing off supervisors' stereotypes regarding the superior work of Asians over Mexicans. The incident was precipitated when a young Mexicana, newly assigned to an assembly unit in which a new circuit board was being assembled, fell behind in her quota. The supervisor berated her with racial slurs about Mexicans' "laziness" and "stupidity" and told her to sit next to and "watch the Orientals." As a group, the Asian women she was stationed next to slowed down their production, thereby setting the average quota on the new boards at a slower than usual pace. The women were in fits of laughter after work because the supervisor had assumed that the speed set by the Asians was the fastest possible, since they were the "best" workers.

Hispanic workers also turn management's anti-Mexican prejudices against them, as a Salvadorean woman explained:

> First of all, the bosses think everyone from Latin American is Mexican, and they think all Mexicans are dumb. So, whenever they try to speed up production, or give us something we don't want to do, we just act dumb. It's not as if you act smart and you get a promotion or a bonus anyway.

A Mexicana operative confided, "They [management] assume we don't understand much English, but we understand when we want to."

A Chinese woman, who was under five feet tall and who identified her age by saying she was a "grandmother," laughingly told how she had her white male supervisor "wrapped around [her] finger." She consciously played into his stereotype that Asian women are small, timid, and obedient by frequently smiling at and bowing to him and doing her job carefully. But when she had a special need, to take a day or a few hours off, for example, she would put on her best guileless, ingratiating look and, full of apologies, usually obtained it. She also served as a voice for coworkers whom the supervisor considered more abrasive. On one occasion, when three white women in her unit complained about poor lighting and headaches, the supervisor became irritated and did not respond to their complaint. Later that week the Chinese "grandmother" approached him, saying that she was concerned that poor lighting was limiting the workers' productivity. The lighting was quickly improved. This incident illustrates that managers can and do respond to workers' demands when they result in increased productivity.

Some workers see strategies to improve and control their work processes and environments as contradictory and as "Uncle Tomming." Two friends, both Filipinas, debated this issue. One argued that "acting like a China doll" only reinforced white employers' stereotypes, while the other said that countering the stereotype would not change their situation, so they might as well use the stereotype to their advantage. The same analysis applies to women workers who consciously encourage male managers to view women as different from men in their abilities and characteristics. For women and minority workers, the need for short-term gains and benefits and for long-term equal treatment is a constant contradiction. And for the majority of workers, short-term tactics are unlikely to result in long-term equality.

Potential for Organizing

Obviously, the lesson here for organizing is contradictory. Testimonies such as the ones given in these pages clearly document that immigrant women are not docile, servile people who always follow orders, as many employers interviewed for this study claimed. Orchestrating major actions such as family migration so that they could take control of and better their lives has helped these women develop leadership and survival skills. Because of these qualities, many of the women I interviewed struck me as potentially effective labor and community organizers and rank-and-file leaders. Yet almost none of them were interested in collective organizing, because of time limitations and family constraints and because of their lack of confidence in labor unions, the feminist movement, and community organizations. Many were simply too worn out from trying to make ends meet and caring for their families. And for some, the level of inequality and exploitation on the shop floor did not seem that bad, compared to their past experiences. A Salvadorean woman I interviewed exemplified this predicament. Her job as a solderer required her to work with a microscope all day, causing her to develop severe eye and back strain. Although she was losing her eyesight and went home exhausted after working overtime, she told me she was still very happy to be in the United States and very grateful to her employer. "I have nothing to complain about," she told me. "It is such a luxury to know that when I go home all of my children will still be alive." After losing two sons to government-backed terrorist death squads in El Salvador, her work life in Silicon Valley was indeed an improvement.

Nonetheless, their past torment does not reduce the job insecurity, poor working conditions, pay inequality, and discrimination so many immigrant workers in Silicon Valley experience in their jobs. In fact, as informants' testimonies suggest, in many cases, past hardships have rendered them less likely to organize collectively. At the same time, individual acts of resistance do not succeed on their own in changing the structured inequality of the division of labor. Most of these actions remain at the agitation level and lack the coordination needed to give workers real bargaining power. And, as mentioned, individual strategies that workers have devised can be contradictory. Simultaneous to winning short-run victories, they can also reinforce both gender and racial stereotypes in the long run. Further, because many of these victories are isolated and individual, they can often be divisive. For workers to gain both greater workplace control *and* combat sexism and racism, organized *collective* strategies hold greater possibilities.

Neither organized labor nor feminist or immigrant community organizations have prioritized the needs of Silicon Valley's immigrant women workers.[6] As of 1989, for example, not a single full-time paid labor union organizer was assigned to the local high-tech industry. Given that Silicon Valley is the center of the largest and fastest-growing manufacturing industry in the country, this is, as one long-time local organizer, Mike Eisenscher, put it, "a frightening condemnation of the labor movement" (1987). That union leadership has also failed to mark for attention a workforce that is dominated by women of color is equally disheartening.

My findings indicate that Silicon Valley's immigrant women workers have a great deal to gain from organizing, but also a great deal to contribute. They have their numeric strength, but also a wealth of creativity, insight, and experience that could be a shot in the arm to the stagnating national labor movement. They also have a great deal to teach—and learn from—feminist and ethnic community movements. But until these or new alternative movements learn to speak and listen to these women, the women will continue to struggle on their own, individually and in small groups. In their struggle for better jobs and better lives, one of the most effective tactics they have is their own resourcefulness in manipulating management's "own logic against them."

NOTES

1. For a comprehensive analytical description of the development of Silicon Valley as a region and an industry, see Saxenian 1981.

2. The production jobs include the following U.S. Department of Labor occupational titles: semiconductor processor; semiconductor assembler; electronics assembler; and electronics tester. Entry-level wages for these jobs in Silicon Valley in 1984 were $4.00 to $5.50; wages for workers with one to two years or more experience were $5.50 to $8.00 an hour, with testers sometimes earning up to $9.50.

3. "Minority" is the term used by the California Employment Development Department and the U.S. Department of Labor publications in reference to people of color. The statistics do not distinguish between immigrants and nonimmigrants within racial and ethnic groupings.

4. In North Carolina's Research Triangle, for example, Blacks account for most minority employment, whereas in Albuquerque and Texas. Hispanics provide the bulk of the production labor force. Silicon Valley has perhaps the most racially diverse production force, although Hispanics—both immigrant and nonimmigrant—still account for the majority.

5. Quality circles are introduced by management for the stated goals of increasing both employee satisfaction and production efficiency by giving workers input into decision-making processes.

6. One of the few organizations that *have* included immigrant women workers and that addresses their needs is the Silicon Valley Toxics Coalition. This group effectively addresses itself to improving residential and occupational health and safety hazards posed by the highly toxic local high-tech industry.

WORKS CITED

Annual Planning Information: San Jose Standard Metropolitan Statistical Area, 1983–1984. Sacramento: California Department of Employment Development, 1983.

Brownmiller, Susan. *Femininity*. New York: Simon and Schuster, 1984.

Carey, Pete. "Tomorrow's Robots: A Revolution at Work." *San Jose Mercury News*, February 8–11, 1984.

Eisenscher, Mike. "Organizing the Shop in Electronics." Paper presented at the West Coast Marxist Scholars Conference, Berkeley, California, November 14, 1987.

Elson, Diane, and Ruth Pearson. "Nimble Fingers Make Cheap Workers: An Analysis of Women's Employment in Third World Export Manufacturing." *Feminist Review*, spring 1981, 87–107.

Fernandez-Kelly, Maria Patricia. *For We Are Sold: I and My People: Women and Industry in Mexico's Frontier*. Albany: State University of New York Press, 1983.

Grossman, Rachel. "Women's Place in the Integrated Circuit." *Southeast Asia Chronicle 6 — Pacific Research* 9 (1979): 2–17.

Hartmann, Heidi. "Capitalism, Patriarchy, and Job Segregation by Sex." In *Women and the Workplace*, ed. Martha Blaxall and Barbara Reagan. Chicago: University of Chicago Press, 1976.

Hossfeld, Karen. "The Triple Shift: Immigrant Women Workers and the Household Division of Labor in Silicon Valley." Paper presented at the annual meeting of the American Sociological Association, Atlanta, 1988.

Lim, Linda. *Workers in Multinational Corporations: The Case of the Electronics Industry in Malaysia and Singapore*. Michigan Occasional Papers in Women's Studies, no. 9, Ann Arbor: University of Michigan Press, 1978.

Nash, June, and Maria Patricia Fernandez-Kelly, eds. *Women, Men and the International Division of Labor*. Albany: State University of New York Press, 1983.

Ong, Aihwa. "Industrialization and Prostitution in Southeast Asia." *Southeast Asia Chronicle* 96 (1985): 2–6.

Safa, Helen. "Runaway Shops and Female Employment: The Search for Cheap Labor." *Signs* 7 (1981): 418–33.

Saxenian, Annalee. *Silicon Chips and Spatial Structure: The Industrial Basis of Urbanization in Santa Clara County, California*. Working Paper no. 345. (Berkeley: Institute of Urban and Regional Planning, University of California, 1981).

Siegel, Lenny. "Delicate Bonds: The Global Semiconductor Industry." *Pacific Research* 11 (1980): 1.

Siegel, Lenny, and Herb Borock. *Background Report on Silicon Valley*. Prepared for the U.S. Commission on Civil Rights. Mountain View, CA: Pacific Studies Center, 1982.

Westwood, Sallie. *All Day, Every Day: Factory and Family in the Making of Women's Lives*. Champaign: University of Illinois Press, 1985.

Chapter Three

Net-Working
The Online Cultural Entrepreneur

Andrew Ross Interviews
McLean Mashingaidze Greaves

While spectacular stories about IPO sensations and digital millionaires abound, the sobering reality is that much like the lottery, the chances of hitting it big in e-commerce and high-tech business ventures are slim. Those chances are slimmer still for an African American entrepreneur trying to balance the need for a greater diversity of content and communities with a profit-making imperative.

In this interview, McLean Mashingaidze Greaves discusses opportunities and limits of online entrepreneurship. Greaves founded cafelosnegroes.com, a vibrant and important "virtual hangout" for people of color, long before racial/ethnic content was considered profitable. Here he talks with Andrew Ross, who has written extensively on economic inequities in the new media work sector, about his trajectory as a black cultural producer.

In this dialogue, conducted online in the summer of 1999, Greaves and Ross explore the following questions: What can be achieved by the inclusion/proliferation of racial/ethnic content sites? Are such sites emblematic of more democratic participation or are they simply new venues for consumption? What does the popularity of cafelos negroes.com and similar sites say about how race works in cyberspace?

Andrew Ross: For those looking for the next dollar bonanza, the Internet of late has been a turbocharged engine of wealth creation for

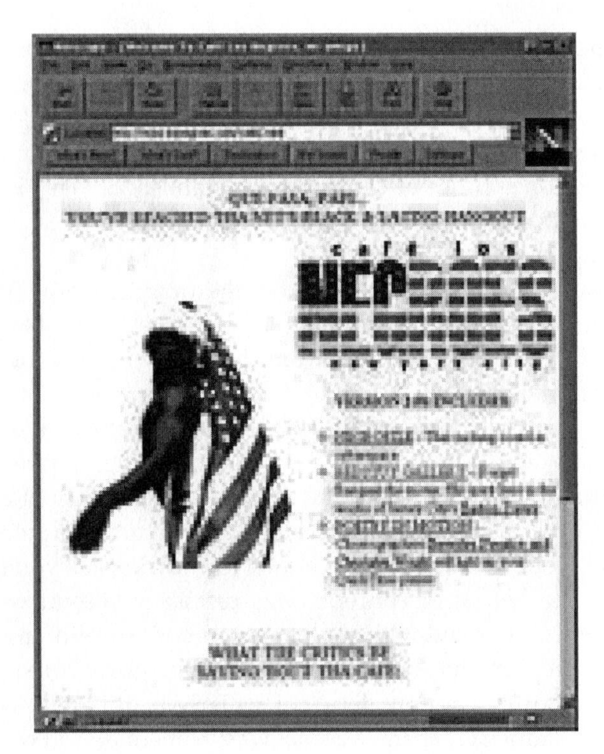

www.cafelosnegroes.com. McLean Mashingaidze Greaves.

WWW entrepreneurs. When you started cafelosnegroes.com in 1996, on the periphery of New York's Silicon Alley, was the culture of greed a little more subdued? Were your goals different from many of your peers who, even then, were looking at start-ups with which they could go public or sell off to the big fish after a few years? Can you describe that early environment for us and where you, as owner of a Web site development company, located yourself?

McLean Mashingaidze Greaves: Actually, the goal behind cafelosnegroes.com and the parent company I founded (Virtual Melanin, Inc.) was to become a market leader and subsequently generate profits. As one of the few black-owned technology companies, however, I found there was a social element from the start, manifested primarily via our various "outreach" programs and internship opportunities.

In the early days of new media, however, the culture was more experimental. At the time, there were no proven revenue models and as such, everything was fair game. My strategy was to maintain brand recognition until the market matured and then go for major financing, once proof of concept was established. (This is almost a prerequisite for black Web firms, which are considered higher-risk for prospective investors.)

Andrew Ross: What form, exactly, did the outreach programs take? And can you clarify how the "proof of concept" double standard worked? Was it any different from the rules of the game in other, nonelectronic media?

McLean Mashingaidze Greaves: We created and taught new media curriculum to high school kids in the Bronx (Wings Academy High School) and Brooklyn (Benjamin Banneker High School and P.S./I.S. 332). We also offered internships and actually provided some contract work to some black/Latino New Yorkers as young as seventeen. The "proof of concept" was primarily self-imposed. As a person who's worked with companies on both sides of the business fence (entrepreneur/venture capitalist), I knew we'd have to have some kind of revenue model—particularly being a black firm in a fledgling industry. There was no way someone would have just thrown a seven-figure check to some dreadlocked computer headz located in a wretched Bed-Stuy, Brooklyn, neighborhood.

Andrew Ross: cafelosnegroes.com, as I remember it, was in part an online arts magazine, and in part a virtual meeting place, but it was also a flag of black and Latino visibility on the Web. Does that sound right? How did people use the site, and why was it discontinued?

McLean Mashingaidze Greaves: cafelosnegroes.com was intended to be a showcase site for black and Latino underground artists. My goal was to provide an editorial voice for the Gen X niche of urban America that was hitherto underserved by the likes of *Ebony, Jet,* and *Essence.* The site also provided an online community for the exchange of news, views, and other information. I closed the site after I tied the knot and decided to end my self-imposed life in the inner city and return to the middle class. The entire experience was essentially a one-man band masquerading as a cutting-edge company. In reality, it was exhausting to work in a squalid environment with error-prone, inexperienced management. Getting married underscored the fact it was time to move on, so I closed CLN after my bid to raise sufficient membership revenue fell short.

Andrew Ross: You sound a little jaded. What were the high points, for you, of the online community that the cafelosnegroes.com fostered? Despite all the hype about the brave new world of Internet publishing, the most common experience of the pioneer Silicon Alley start-ups was one of crushing overwork and financial frustration. Do you think that cafelosnegroes.com faced any out-of-the-ordinary obstacles?

McLean Mashingaidze Greaves: I'm probably a little jaded because it was really just a matter of great vision, poor management. I can't begin to tell you all the crazy, behind-the-scenes screwups that occurred. That's where having enough capital to put together a solid team (as opposed to a ragtag band of "volunteers") really mattered. Unlike our white counterparts, entrepreneurial and technology skills are not widespread in the black community. Plus, we haven't been middle-class long enough to have a tradition of high-risk entrepreneurialism. That dilemma was the key obstacle we faced at cafelosnegroes.com. The high point, to be honest, was getting rid of the stress of working under such conditions. On some levels, it was an interesting, educational experience but it usually felt like doing jail time.

Andrew Ross: You mention that CLN was intended to cater to "the Gen X niche of urban America that was hitherto underserved by the likes of *Ebony, Jet,* and *Essence.*" Yet, the glut of culture consumer magazines in the *Vibe, Blaze, Source* category targets minority youth at market saturation levels. The online world is another matter, and surely was much more so when CLN started up. What other online sources were there at that time? And what did you yourself discover about the tastes and opinions of the Gen X niche of urban America in the course of running CLN?

McLean Mashingaidze Greaves: At the time I created CLN, the only other well-known urban sites were Net Noir, Universal Black Pages, and New York Online. The latter was in its last days. Once we added interactive items like message forums, a strong online personality developed. Immediately I noticed our audience had a strong female voice (55 percent of our audience), was very political, and had a strong artistic bent (as evidenced by our thriving poetry board). The CLN audience understood the "inside" jokes in my editorial stance and had an appreciation of our undiluted pseudo-intellectual Afro-isms. It was interesting to see a number of users bonding offline. Some of them came up with some CLN spin-offs including a user calendar (with photos) and member content like the now-defunct bwitch.com. The entire thing was created without a marketing budget or corporate prodding.

Andrew Ross: Not many of the revenue models dreamed up by the pioneer online publications came to fruition. Those publications that survived tended to hitch themselves to much bigger corporate outfits. Are there any lessons you have learned about the unholy marriage of culture and commerce that might have been applied, in retrospect, and that would have made a difference?

McLean Mashingaidze Greaves: To be honest, the key thing I learned was the importance of adequate capital and subsequently landing a top-notch management team. In the cut-throat, fast-moving business of new media, a tiny start-up (particularly one located in a high-crime, underclass neighborhood) is only as strong as the weakest link. In retrospect, it would've been better to partner with a deep-pocketed corporate backer and assemble the best that money could buy. In urban new media, it's all about bucks, building, and branding—in that order.

Andrew Ross: cafelosnegroes.com was founded just after the peak of the first wave of a black arts renaissance in New York and other cities. You might have expected a good deal of the energy from these movements to flow online. Was there any evidence of this, or did you come across any resistance to the new medium from those blowing up in the traditional arts media?

McLean Mashingaidze Greaves: We found a significant demand for undiscovered talent on CLN. My general focus was to give exposure to artists who were on the verge of blowing up in their careers, thus establishing editorial credibility for having an ear to the street. Many of the artists we covered went on to greater success. The online audience generally posted support toward these artists, often sharing info about their favorites. Without question we tapped into an online movement that stemmed from the post–civil Rights, Gen X demographic one would find in neighborhoods like Brooklyn's Fort Greene/Clinton Hill or the underground loft party scene in Oakland, California.

Andrew Ross: You also attracted some high-profile clients, some of them household names, for your Web design services, while CLN generated a pile of press clippings in no time at all. The workings of online fame, like Internet capitalism and digital machinery in general, seem to operate a lot faster than we are used to seeing. What happened in your case? And do you believe these speed upgrades amount to significant changes in the way that culture is transacted in late capitalist society?

McLean Mashingaidze Greaves: Interesting question. When I started CLN, it was pretty easy for me—as a journalist/entertainment biz professional—to figure out there weren't a lot of media stories involving blacks with technology at the time. The high-risk notion of a black male quitting a full-time, cozy job as a magazine editor in Soho to start an unprecedented black high-tech firm in a decrepit inner-city neighborhood was newsworthy on a number of levels. Hence the avalanche of press. There still isn't an African American media equivalent of Steve Jobs or Bill Gates although it's inevitable that we'll eventually produce black cultural heroes known more for depth than dunks.

In my case, the path was very clear but, as with Jobs or Gates, technological prowess must be supported with profits. Since CLN never became a box-office hit like AOL, my online fame has more or less hovered at the underground level, although I attract roughly half a dozen articles annually and several academic opportunities. Unlike the real-world version, online fame doesn't come with the perks to justify the crap. In fact, since cyberspace puts the talent and audience on the same interactive level (server-side controls notwithstanding), crazed, obsessive fan behavior comes through louder, clearer, and more pervasive than before. Conversely, the fame part doesn't offer perks like seven-digit paydays, personal security, and exclusivity.

I think the Net will eventually offer lucrative celebrity opportunities. Already, fifteen-year-olds are selling homegrown Web sites to large content companies at million-plus prices. As bandwidth increases, it'll be much easier for "thinking" celebrities to establish a meaningful level of fame and thereby attract revenue without having to sacrifice anything beyond time. The barrier of entry will remain low for a long time and, if anything, the process will be incredibly simpler with a bigger audience to boot.

Andrew Ross: Entrepreneurial investment and energy in general seem to be a high priority for you, and yet CLN and its successor Digital Downlow tap into the mentality of urban bohemians with underground loyalties, in other words, a mentality that is traditionally allergic to the language of venture capital. How do you reconcile this apparent contradiction?

McLean Mashingaidze Greaves: There is an ostensible contradiction that is just that—ostensible. Underground "bohemians" with Net access generally are middle-class and thus have personal disposable incomes to spend on nonessential items like Net access. The reason I've focused on a very specific audience is that content providers must dominate niche markets in order to establish a presence where viewers literally have millions of channels to surf. At the time of CLN, this market was too small to make any serious e-commerce waves. These days, the word "urban" is used by black Web sites that wish to maximize ad revenue. Like traditional broadcasting, the term implies that crossover appeal is needed to attain critical mass. Even more so, with the Web. As a result, commercial black Web sites will target a centrist editorial approach, thereby continuing the *Essence/Ebony* tradition of "safe" content. It's just a matter of time before the post–Civil Rights, Gen Xfro crowd I targeted becomes large enough to demand a fresher journalistic style.

Andrew Ross: Readers may be surprised to learn that the actual physical location of an online company—Bed-Stuy (a low-income Brooklyn neighborhood), in the case of CLN—makes a real difference to its ability to attract investors. The given wisdom about the wired world is that physical place no longer matters. It sure does, doesn't it? People go to work, and cyberspace is manufactured in these workplaces, like any other product.

McLean Mashingaidze Greaves: The real workplace is very much where products are manufactured and meetings are held. I was always amazed to see white, corporate types show up at a soot-stained, non-AC, unkempt apartment office in a rundown Brooklyn building (where it wasn't unusual to see drug deals go down in front of the "office entrance") in the name of new media. The free market system produces peculiar scenarios.

Andrew Ross: Aside from the verifiable existence of Gen X (or Gen Xfro) as a market demographic, what did you learn about the cultural and political instincts of your online audience, as distinct, say, from the values of a Civil Rights generation?

McLean Mashingaidze Greaves: Our online audience was more or less a reflection of my own generational values (the editorial commonality that generated site traffic). Those values placed special emphasis on pluralism, upward mobility, and, ultimately, independence. Culturally, there was probably more intellectual diversity in cafelosnegroes.com than you'd ever find in the entire catalog of black television. Politically, our audience was skeptical of the mainstream establishment, but that viewpoint isn't particularly groundbreaking in any predominantly African American environment, be it virtual or inner-city.

Andrew Ross: Life after CLN. Your company VMI has developed, or is developing, a range of sites: Next Step in Philadelphia, Negrofile, Afro-Picks, and the Digital Downlow, the successor, in many ways, to CLN. Tell us about the reasoning behind these particular projects.

McLean Mashingaidze Greaves: Ultimately the goal behind those sites is to nurture a roster of "best of breed" sites and aggregate them in a leveraged manner à la Lycos. Presently, however, I am more involved with strategies for porting television content onto the Net with an eye toward entertainment (as opposed to promotional or informational).[1] In my work as a consultant, my clients already include media giants like HBO, Time Warner, and CBC (the Canadian Broadcasting Corporation). It's a kind of preparation/self-training for the next big development in the Internet business—high broadband.

Andrew Ross: The Digital Downlow hosts several message forums: the Brawley Room, for discussion "on race relations à la Tawana Brawley"; the Dread Poets Society, for underground griots; Q-Tip, "Gay/Lesbian issues from a colored perspective"; Tech Headz, "technology and futurism from a Gen Xfro viewpoint"; Campus Chats, for student life; BeatNet Beats, for music heads; Foundation for Ethnic Understanding, on Black-Jewish relations; and a general Digital Downlow salon. Why these? Were they the result of formal, or informal, market research? How do you measure the success of each of them? By traffic volume, or by the quality and genuine dialogue quotient of the postings?

McLean Mashingaidze Greaves: As I did with most of our original content products, I essentially created the editorial, graphic, and HTML design on the fly, as quickly as possible. As such, the categories for our message forums were a little arbitrary, but I also tried to focus on areas that addressed key demographics within our audience. The long-term goal—like the rest of our content—was to build mini-brands within the site. Based on the number of posts and subsequent file sizes, our most popular CLN salon was Digital Downlow—a free-for-all, no-holds-barred discussion board. After that, the music and poetry sections maintained very strong, loyal followings. Naturally, the traffic served as hints for additional potential niche sites.

Andrew Ross: How many of your original CLN subscribers/participants has the new site retained? Were any of them clamoring for a successor, and did any of these folks have any input into the construction and content of Digital Downlow?

McLean Mashingaidze Greaves: The Digital Downlow site has retained a majority of our original hard-core CLN members as well as an audience unfamiliar with the old content. It was essentially created to satisfy the needs of the original community who felt they would not be able to find the same kind of audience/environment elsewhere on the Web. To be honest, the site is still very much in beta.

Andrew Ross: As you know, the rise of the Internet has been accompanied by a torrent of brave new world rhetoric, whether tied to the promise of universal access ("Log On and You Will Be Free") or to the potential for unfettered speech ("The Virtual Republic"). Much of this is crass hype, cranked out to feed the expansionist fantasies of the telecom Goliaths and other neoliberal enthusiasts. On the global stage, it's a sobering thought that a huge percentage of the world's population will live and die without ever having made or received a phone call. And it remains the case that access to income is a priority that dwarfs all talk about access to information technology. On the other hand, there's every reason to make serious attempts to rectify the limited media access accorded to populations and communities that, historically, have been socially denied and disenfranchised. Where do you see the solid potential of such efforts, and how do you separate it from the snake oil hype?

McLean Mashingaidze Greaves: In my view, it's in the interests of capitalism to see the online market expand to the fullest extent possible. It's also in the interest of government to have such a powerful surveillance option in place. That said, I strongly believe that the economic thrust of new media will force a trickle-down effect wherein technology will become ubiquitous—even among the socially disenfranchised.

The key difference in the digital revolution won't be so much an issue of download access as it will be an issue of upload. This means that typically consumerist, socially denied communities will easily continue that role in the new medium as "download" audience members (à la mouse potatoes) via low-cost, user-friendly appliances like WebTV. In contrast, this growing market will support the producers of online content who, in my view, will benefit in an "upload" sense—meaning they will be the ones profiting from uploading and subsequently delivering content to the aforementioned downloaders. Nowhere will this new reality be more obvious than the eventual divide between quality bandwidth (HDTV and Internet2) and grassroots bandwidth (the current Internet).

Andrew Ross: One of the mythologies about the Internet holds that it is a relatively rule-free environment of informal protocols, yet it is already highly standardized. Standardization tends to occur quickly in any new technological environment, whether imposed by commercial monopoly or by publicly minded bodies, and it usually excludes those with fewer resources or with bad manners. The recent debates around ICANN (Internet Corporation for Assigned Names and Numbers), which controls the Internet's root name servers, and NSI (Network Solutions Incorporated), a monopoly since 1993, relating to competition in the domain name registration market, may be warning signs that the present skirmishes about the commercial value of intellectual property will quickly escalate into battles about the governance of the Internet. Given the perceived need, increasingly, for regulatory bodies in new media, where do you think those authorities are best located among corporations, nonprofit institutions, or government agencies? And whose interests should they serve?

McLean Mashingaidze Greaves: In my egalitarian view, the Net should be governed by a self-regulating body democratically elected by Netizens around the world, using secured polling technology. Of course, such a notion flies in the face of capitalism and thus will not occur. The second best option would be to allow a nonprofit organization complete privilege to control root servers et al. Without question, the governing agency should serve the interests of both copyright holders, merchants, and developers as well as e-commerce consumers and other online users.

Andrew Ross: One of my own research interests has been in new media labor patterns, not just among Silicon Alley programmers, designers, and coders, who (with the exception of some whiz kid designer poster-employees) are overworked and underpaid in nonunion environments, but also among the armies of data processors in places like Bangalore and the Caribbean. In addition, I have found that the industry's concern for the labor conditions of perma-temps does not extend to the workers in hardware manufacturing. Workers in semiconductor plants have the highest rates of industrial illness; these plants cause some of the worst environmental impacts of any form of manufacturing. They are mainly located in or near low-income communities of color in the United States or overseas, and increasingly resemble the sweatshop economy of the garment industry. Why do people in new media turn a blind eye to such appalling labor conditions? Unlike fashion professionals, who know all about sweatshops and try to ignore their existence, most new media professionals don't seem to know very much at all about how their technologies are made.

McLean Mashingaidze Greaves: To be honest, most new media professionals are too overwhelmed with their own personal manifest destiny in this frenzied industry that everything in the nonwired world becomes relatively abstract. It's important to note that most Internet workers are young, Gen X types who are experiencing the first (and likely only) gold rush of their generation, usually from the viewpoint of their own aesthetically pleasing sweatshops while they wait for their stock options to vest. As such, the technology purely becomes a means to an end. In this fast-paced business, who has time to actually consider the origin of the highly disposable, increasingly obsolete tools of the trade?

Personally, I don't remember *ever* working a five-day week. Almost everyone I know in the technology business works at least six days a week. The notion of "appalling labor conditions" is less relevant when work is entirely portable. Instead, the imminent dilemma involves a never-ending work schedule and the toxic effects of its "trojan" nature.

Andrew Ross: Cultural critics like to discuss what difference "race" makes to modes of human expression. Because it offers relatively disembodied contact, there are commentators who see new media as some kind of race-less medium. Others are less inclined to view the development of this technology, and its impact, as a color-blind process. In your opinion and experience, how does race matter in cyberspace?

McLean Mashingaidze Greaves: Yes, despite the various ironies, race does matter in cyberspace. It matters with access (where African Americans are falling behind for various reasons). It matters with content (where there is still no real entertainment equivalent of Harlem in the 1930s and 1940s or even Bill Cosby circa 1985).

Most importantly perhaps, race matters in cyberspace where economics are concerned. A veritable "gig rush" is going on right now with all kinds of speculators staking claims on the digital economy. Like the land grab by early American settlers, there's a lot at stake here and too many people of color are being left out of the billion-dollar valuations and IPOs that are almost becoming commodities unto themselves.

Finally, the networking nature of the Net lends itself to communities where the common denominator is often fundamental similarities like sexual orientation, hobbies, and even race. Despite its digital interface, anonymity on the Net is generally transitional where online communities are concerned. As such, the ultimate objective of networking demands self-revelation at some point.

Then again, as ethnic power users know, free your mind and your ASCII will follow.

NOTE

1. "Porting" is a computer term for transferring data between platforms (as in "porting an application to a different operating system").

Chapter Four

Temporary Access
The Indian H-1B Worker in the United States

Amitava Kumar

> Masses of people work in cyberspace or work to make cyberspace possible.
>
> —Andrew Ross

> We'll only achieve our full potential if we assure that high-tech companies can find and hire the rare people whose skills are critical to America's success.
>
> —Senator Phil Gramm

Writing Codes

There are (at least) two narrative lines about work and the new wired order.

One can find an example of the first narrative in an editorial about the influence of information technology on the workplace in the *Economist*. According to this editorial, "It is a cliché to say that 'the *Internet changes everything*': the challenge now is to guess what, how, and how quickly."[1] One of the more significant changes has been in the structure of the workplace. To make it easy for us to imagine this change, the editorial draws a comparison with the transformation of the big studio system in Hollywood: "Once, a Hollywood studio employed everyone from Humphrey Bogart to the lighting technicians."[2] Today, studios have retreated from their traditional roles as employers and now assemble the teams of self-employed people and small businesses that are today's stars and technical support. Now, everyone's a free agent.

Like Hollywood, large First World companies increasingly rely on outsourcing and communication technologies to foster business relations with suppliers and distributors. With the aid of information technologies, capital now conducts its romance strictly on the basis of ad hoc partnerships and alliances; fluid and mobile contracting firms deliver goods and services to a well-connected world. The difficult questions about the consequences of this change, particularly for the most vulnerable parties in this process, are glossed over. Instead, this narrative stresses the magic of change.

A second narrative can be found in *Little India*, an Indian newspaper published in the United States. In these pages, there has been much discussion about the existence of what have been called "bodyshops"—contracting firms, mostly owned by Indians settled in the United States, that hire H-1B workers from India.[3] An essay in *Little India* entitled "The Bloodsuckers" began with the words, "Let me relate the experience of a typical victim—we can't give him a better name than Mr. Oracle Rao." Oracle Rao in this narrative is a composite of the varied experiences of Indian H-1B workers. The story of Oracle Rao, a newly arrived Indian programmer, depicts both the anticipation and disappointment that typify the H-1B experience.

Shortly after arriving, Rao learns that he is being charged for food, lodging, and transport but that he does not yet have a job. He undergoes interviews for jobs that "could be anywhere from California to Maine or Texas."[4] Rao may be moved at the whim of the client. Or he may be subcontracted to another firm, in which case his earnings will be shared by the two firms that have now invested in him. He may not get a job at all; in the lingo of the trade, he could be "benched." During this period, it is possible that he will not be paid—but at the same time, he will continue making payments for his expenses to the firm that brought him here. More than the itinerancy, as I found out in the course of the interviews I conducted, it is the uncertainties of the period of "benching" and its attendant humiliations that most burden H-1B workers' minds.[5] But all of these humiliations are balanced by the hope of making more money and having a better life in the United States.[6]

It is this second narrative that holds my attention. Not so much because the *Little India* story is more true than the one hawked by the

Economist, but because the only foreign-sounding names in the latter's report are the names of computer languages. In other words, race and geography have been deleted from the *Economist*'s extensive update on e-business, to leave dangling the question of the so-called Third World in the latest developments in technology.

But the mainstream press in the United States has not been wholly silent on the issues of race and geography. Among the earliest reports on these issues was one entitled "White-Collar Visas: Importing Needed Skills or Cheap Labor?" in the *Washington Post* in 1995. The article began with this alarmist news: "A large New York insurance company [laid] off 250 computer programmers in three states and replace[d] them with lower-wage temporary workers from India."[7] The subtle shift from "computer programmers" to "temporary workers" is a veiled statement that H-1B programmers from India have lower skills and are willing to accept lower wages. In fact, this article quotes Lawrence Richards, a former IBM computer programmer and founder of the Software Professionals' Action Committee, as saying that the H-1B system is being used "not to obtain unique skills, but cheap, compliant labor." Richards describes the imported professionals as "techno-braceros," the high-tech equivalent of migrant farm workers. It shouldn't be taken lightly that these H-1B workers have been paralleled to other racial groups whose labors have also been deemed cheap and who share historically vulnerable positions in the workforce.

While mainstream accounts like those in the *Washington Post* do consider the larger issues of race and nation, they fail to consider the impact of these changes and policies on the lived experiences of Indian H-1B workers. Clearly, Richards is unsympathetic to those workers, whom he has described as being "indentured" to their employers; indeed, he started his action group to advocate for his colleagues replaced by Indian programmers willing to work for lower wages. What Richards failed to see was that immigrant technology workers also bear the costs of employment restructuring that are represented by the character of Oracle Rao.

One real-life example of the Oracle Rao phenomenon is the story of Satyajit Roy. Roy is a thirty-nine-year-old software engineer who works for a large telecom company. Before coming to the United States he had several years' experience as a manager in the technology field in India. He is on his fourth H-1B visa.

A small bodyshop had given him the chance to come to the United States, but it couldn't very quickly find a place for him. "In the very beginning," Roy said, "I didn't have a car, I didn't have a salary." His present job is the first he has found satisfactory since his arrival in this country from India two and a half years ago. Recently, in his new job, Roy's boss has promised to create a permanent position for him. Roy is looking two or three years ahead when he hopes he will have a green card.[8]

Roy is a little different from most Indian cybertechies, who usually come earlier in their careers and alone. He moved here with his wife, Debjani, and their son. When asked what had motivated him to come here, he replied, "I wanted my family to get more exposure to the world. I had been in England when I was younger. I wanted to give my family that same experience."

When I first met Roy, we spoke about the India and Australia World Cup cricket match being played in England. Roy had seen parts of it on television at work with his cohort of H-1B coworkers from India. "The H-1B visa is a big boon to Indians, as I see it," Roy said to me. "From the mid-eighties, the efflux really started, and now, we are a class by ourselves."

What are the conditions that have created this class of H-1B workers? Much of the vaunted success of the Indian cybertechie was spurred in the early 1990s, when U.S. companies hired Indian programmers and consultants in droves. India is overwhelmingly the largest supplier of information technology (IT) professionals to the United States. When the annual cap on H-1B visas was raised from 65,000 to 115,000 in 1998, Indian cybertechies filled 46 percent of that new total. China filled 10 percent of that number; other countries with significant rates of labor immigration to the United States include Canada with 4 percent, the Philippines with 3 percent, the United Kingdom, Taiwan, Pakistan, Korea, Russia, and Japan with 2 percent each.[9] As a result of the jitters over the Y2K millennium bug, the demand for IT professionals from abroad became so great that several leading U.S. technology companies lobbied the federal government to increase the H-1B visa cap. In response, Senator Phil Gramm recommended that the cap be raised to 200,000 by the year 2000, arguing that H-1B workers were "rare people whose skills are critical to America's success."[10]

Girish Bhatt, an executive at the Chicago-based CyberTech Systems,

Inc., with a staff of over five hundred H-1B workers, told me in a phone interview that "the big boom has come to an end, especially in ERP [enterprise resource planning]."[11] An H-1B worker, Sandeep Singh, also based in Chicago, added a twist to this scenario: "Everybody is scared about Y2K, so no one is starting a new project."[12] In the past year, said Singh, the demand for H-1B workers has died down and there has been a slump in hirings. Girish Bhatt concurs: "Those workers who came here to provide solutions to the Y2K crisis, and who have not had an opportunity to learn new skills, now face a dead end."[13] Now, the Indian cybertechie will need to have more than a basic knowledge of software in order to survive or prosper in the new millennium.

The Binary Scene

In a short story by Vikram Chandra entitled "Artha," we find not one, but two rarities of Indian fiction: gay lovers and computer programmers. To make matters more complex, the gay lovers are Hindu and Muslim, and Sandhya, the main programmer in the story, is female. This is how the narrator explains the difference between his and Sandhya's work to his male lover:

> I put my hand on the back of his hip, with a finger looped through a belt hoop, and told him again that I coded high and she coded low, that when I cranked out my bread-and-butter xBase database rubbish I was shielded from the machine by layers and layers of metaphor, while she went down, down toward the hardware in hundreds of lines of C++ that made my head hurt just to look at them, and then there were the nuggets of assembly language strewn through the app, for speed when it was really important, she said, and in these critical sections it was all gone from me, away from any language I could feel, into some cool place of razor-sharp instructions, "MOV BYTE PTR [BX], 16." But she skated in easy, like she had been born speaking a tongue one step away from binary.[14]

This comment reveals the ways technology, even when it promotes access to a better way of life for many, insinuates into an iniquitous system one more element of oppression.

What is the experience of women who are not engineers, but whose destinies, as a result of marriage, are nevertheless tied to the debates on high-tech immigration? The case of Debjani Roy, the spouse of Satyajit Roy, reveals an often-overlooked dynamic in H-1B immigration. H-1B visas are granted only to the workers themselves; spouses and children who accompany them are given H-4 visas, which allow them to stay but not to work. When a worker is "benched" or a contract is canceled, the family fully shares the brunt of the burden. While Satyajit Roy waited for work, his family waited with him. Debjani, who was an English teacher in India, was forced to quit her job and depend on Satyajit's salary.

R. Mutthuswami notes that in addition to these gender divisions, there is also a "clear dichotomy" between two classes of Indian programmers. One class comprises highly educated Indians "who have given up academic careers to start their own companies" in places like Silicon Valley, and the other is mostly made up of graduates from regional colleges and less prestigious programs who perform "low-level coding jobs in the U.S., Europe, or Australia."[15] Those who fall in the former category, according to Mutthuswami, today serve as the CEOs of 25 percent of the companies in Silicon Valley. The members of the second class are those who form "a larger portion" of Indian cyberworkers in this country. They perform "manual work," Mutthuswami said with a shrug, and added, "It is a class system, like any other class system."

I asked Mutthuswami whether Oracle Rao, our composite H-1B programmer, would have any say in where he was placed in this hierarchy. Speaking broadly of the class of Indian cybertechies on H-1B visas, Mutthuswami said, "No. They don't get paid very well, they don't have any power or clout. They have skills, but they are mostly for maintenance jobs. I see nothing intellectual coming out of their work here." But Muttuswami simplifies what is a very complicated issue. While it is true that Indian programmers receive less compensation than their American-born colleagues for the same work, it is also true that, according to Satyajit Roy, "[n]o one comes here [for] below 40K."[16] Furthermore, it is inaccurate to suggest that they are all low-skilled workers; in fact, "many of them provide sophisticated expertise to most of America's Fortune 500 corporations."[17]

The Indian computer culture is therefore immensely complex. It is divided between transnational capital and workers, men and women, highly skilled and less skilled. Needless to say, it is the workers who, while hoping to gain from the H-1B program and even find permanent residency in the United States, remain most susceptible to the system's injustices.

Rebooting

"India in the 1950s," Sunil Khilnani has written, "fell in love with the idea of concrete."[18] In the 1980s, one might add, India fell in love with the idea of computers. To many this might suggest a massive paradigm shift. But is it?

The champions of market liberalization in India, starting with Rajiv Gandhi in the 1980s, have been advocates of what Sam Pitroda, the first chairman of India's Telecom Commission, calls "software." Pitroda makes a distinction between "software" (education, health, communications technologies) and "hardware" (the concrete dams and steel factories that Nehru had described as the "new temples" of modern India).[19] But the real opposition is not between "software" and "hardware," but between those marginal few who benefit most from these changes and those others who remain either spectators or victims. For me, this binary presents a different set of problems. Whether investing in steel or in silicon, modernity has left untouched—and, in several cases, actually harmed—the lives of many millions of Indians. Sundeep Waslekar has written,

> [A] few million urbanites, white collar workers, trade union leaders, large farmers, blackmarketeers, politicians, police officers, journalists, scholars, stockbrokers, bureaucrats, exporters and tourists can now drink Coke, watch Sony television, operate Hewlett Packard personal computers, drive Suzukis and use Parisian perfumes, while the rest of the people live in anguish.[20]

Our readings of the strange destinies of Oracle Raos must, therefore, be carried out against the pronouncements made by the leaders of government and industry. The materialist reading that I have been calling for is one that finds in the margins of the more abstract claims of the state and capital the smaller, often contestatory, truths of other unheralded lives.

As Andrew Ross reminds us, cyberspace "is not simply a medium for free expression and wealth accumulation; it is a labor-intensive workplace."[21] Ross wants us to pay attention to issues of labor and compensation:

> [T]o focus only on the creative sector of [new media's] contingent workforce is to encourage the assumption that it is wholly disconnected from industrial sites of very cheap labor—electronic chip production and circuit assembly in Asia and the Caribbean, and the armies of word processing and data entry clerks in Ireland and India.[22]

Despite the vulnerability of cyberworkers, Edward Yourdon's *Decline and Fall of the American Programmer* posited a hostile takeover of the U.S. software industry by the likes of Oracle Rao.[23] Yourdon complained that "hardly anybody seems to be paying attention to the fact that a programmer in India earns five times less than a programmer in Indianapolis."[24] He felt that India and other former British colonies posed a serious threat to the United States because they had "inherited an excellent English-based educational infrastructure."[25] In addition, more than 50 percent of the U.S. computer science Ph.D. students were foreign nationals.

Yourdon had an additional point to make: India loses its best software engineers each year to Australia, England, Canada, and the United States.[26] The phenomenon, described so often as the "brain drain," remains the most dramatic and consequential detail in the world of software technology for India. "[B]rains go where the money is. And the money is in technology, and the technology is in the West," notes the British Marxist theorist A. Sivanandan.[27] Yet conditions are not as simple as his comments might suggest. Transformations in technology make divisions of labor increasingly ad hoc, temporary, and fluid. Even Sivanandan recognizes this when he writes that "some of the low-tech, labor-intensive work in the garment industry, which was once being farmed out to the FTZs in Asia, has now, because of new manufacturing techniques combined with cheap female Asian labor [in the United Kingdom], come back to Britain."[28] In other words, the software industry is located in both the First and Third Worlds; in both places, in different modes, global capitalism works to extract surplus from its laborers.

However, the seriousness of the "brain drain" does not deny the very promising growth of the Indian software industry. According to an unpublished report prepared for the Indian government, the industry recorded "a growth of 58% in production and an increase in exports by 76% in 1997–98."[29] According to the government figures, the target for software exports in 1998–99 is $2.6 billion. Additionally, as many as 158 of the Fortune 500 companies either have their own software setups in India or have ties to Indian software companies.[30] To ensure that the industry does not lose this momentum, the Indian government has established a National Task Force on Information Technology and Software Development. The task force hopes that by the year 2008, the annual export of computer software will be worth $50 billion and the export of telecom hardware worth $10 billion. Dewang Mehta, the president of the National Association of Software and Service Companies, India's premier body of computer software and service companies, boldly declares on its Web site,

> I share a dream with my countrymen [*sic*] in India. We all wish to see India emerge as a software superpower. . . . We have to work together in penetrating IT [information technology] to every nook and corner of the country. In this direction, it is also important, say in [the] next five years, to aim for 100% literacy in India.[31]

Does the development of the Indian software industry and the entry of Indian H-1B workers augur a change in the relations of production in the world of cybertechnology? Perhaps not, but their presence and if not their skills then certainly their histories introduce contradictions into the system that are not always easily absorbed or dissolved. Technological experts from a zone of underdevelopment, they introduce an undeniable instability in the dominant imagination. The debates on immigration caps and reports on the huge efflux from India also bring to light an important fact: new developments in cybertechnology have opened avenues of upward mobility for skilled Third World professionals in the United States. Further, the debates around exploitation in computer bodyshops serve as a reminder to those who had all too quickly built their high-tech utopias that the need for structural change cannot be so easily bypassed.

Conclusion

In an essay entitled "Cycles and Circuits of Struggle in High-Technology Capitalism," Nick Witheford has written that "our travels along capital's data highways have discovered at every point insurgencies and revolts, people fighting for freedom from work, creating a 'communications commons,' experimenting with new forms of self-organization, and new relations to the natural world."[32] Many of these issues have surfaced recently in a provocative way with the publication of a study by the Public Policy Institute of California that argues that Chinese and Indian immigrants, who head nearly a quarter of Silicon Valley's high-tech companies, rather than stealing jobs, are actually generating sales of almost $17 billion and providing 58,000 jobs.[33] But others have quickly pointed out that these figures hide a more depressing truth. According to one writer at the *Los Angeles Times*, Charles Piller, the decades-long growth in Silicon Valley has left many behind. "Much of the Latino community, the area's largest and fastest-growing minority group, has seen its standard of living and education deteriorate as the local economy bursts its seams."[34] The dazzling growth experienced in Silicon Valley has sent up housing and other costs in the area so that the low-skilled workforce cannot afford to live there any longer. "In short, a sizable underclass," Piller adds, "has not just been excluded from stock options, it has been harmed by Silicon Valley's overheated wealth creation."

In the end, there is a lesson here also about the technology that has fueled the rise of Silicon Valley. It is precisely the Internet and computers, which were supposed to produce a borderless world, that have thrown up, in the above context, a racially marked sense of the local. Whether or not the Indian H-1B worker is able to program himself or herself to connect with this emergence is not a matter of either technology or politics alone, but both.

NOTES

1. "When Companies Connect," editorial, *Economist*, June 26, 1999, 19 (emphasis added).

2. Ibid., 20.

3. It is under the H-1B provisions of immigration law that "temporary skilled workers" are given visas to enter the United States. Many "bodyshops" have been accused of exploitative practices, including underpayment and harassment; because the H-1B workers are temporary workers, and dependent on such "bodyshops" and other employers for acquiring a

green card when their H-1B visas expire, they remain especially vulnerable to exploitation.

4. Amarnath Vedachalam, "The Bloodsuckers," *Little India*, September 1998.

5. In some cases, the names of the interviewees have been altered at their request.

6. As Vedachalam notes,

> For these Indians life is here better indeed as they make a lot more money than they did in India, especially considering the currency converter. [Bodyshops] use these professionals for making money. They bring them on H1 permits, put them at client sites, charge clients about $8,000 per month, but pay their employees only half as much (including all benefits). Thus the contracting company gets $4,000 per month per employee.

7. William Branigin, "White-Collar Visas: Importing Needed Skills or Cheap Labor?" *Washington Post*, October 21, 1995, A01.

8. The prospect of getting a green card is very limited. Thus, the future employment prospects of many of these workers remain uncertain because the employment-based cap on green cards has remained fixed at 140,000 since 1990.

9. Aziz Haniffa, "Ceiling Is Reached for H-1B Visas," *India Abroad*, June 25, 1999, 38.

10. Quoted in ibid.

11. Enterprise resource planning (ERP) is an application that integrates all aspects of a company's far-flung operations—from receiving through distribution and everything in between—over the company's network.

12. Sandeep Singh, telephone interview by author, June 12, 1999.

13. Girish Bhatt, telephone interview by author, June 19, 1999.

14. Vikram Chandra, *Love and Longing in Bombay* (New York: Little, Brown, 1997), 167.

15. R. Mutthuswami, interview by author, June 8, 1999.

16. Satyajit Roy, interview by author, June 23, 1999.

17. Cyrus Mehta, "Deal Reached to Increase H-1B Visas: Is It Worth It?" *siliconindia*, September 1998, 52.

18. Sunil Khilnani, *The Idea of India* (New York: Farrar, Straus and Giroux, 1998), 61.

19. See http://209.35.112.48/magazine/Aug97sam.html.

20. Quoted in Sashi Tharoor, *India: From Midnight to the Millennium* (New York: Arcade, 1997), 175.

21. Andrew Ross, "Jobs in Cyberspace," in *Real Love: In Pursuit of Cultural Justice* (New York: New York University Press, 1998), 12.

22. Ibid., 20.

23. Thanks to R. Mutthuswami for bringing this text to my attention.

24. Edward Yourdon, *Decline and Fall of the American Programmer* (Englewood Cliffs, NJ: Yourdon Press, 1992), 4.

25. Ibid.

26. According to Yourdon,

> A study by the IIT [Indian Institute of Technology] in Madras found emigration of its graduating class was at the 20 percent level through most of the 1960s and 1970s, but rose to 27 percent in the 1978–82 period, and to 35 percent in the 1983–87 period. Even more dramatic is the figure for graduates in computer science and engineering, a degree program instituted in 1982: of the first batches of graduates in 1986–7, 58.5 percent emigrated. What's the point of training 100,000 new software engineers if they're all going to move away in search of higher salaries?

Ibid., 297.

27. A. Sivanandan, "Heresies and Prophecies: The Social and Political Fallout of the Technological Revolution," in *Cutting Edge: Technology, Information Capitalism and Social Revolution*, ed. Jim Davies, Thomas A. Hirschl, and Michael Stack (New York: Verso, 1997), 290–91.

28. Ibid., 290.

29. Thanks to K. V. Pratap at the Indian Consulate in Washington, D.C., for making this document available to me.

30. Such companies include Coca Cola, British Airways, Swissair, Alcatel, Prudential Insurance, Citicorp, American Express Bank, GE, PepsiCo, Intel, and Informix.

31. See www.dewangmehta.com.

32. Nick Witheford, "Cycles and Circuits of Struggle in High-Technology Capitalism," in Davies, Hirschl, and Stack, *Cutting Edge*, 232.

33. Annalee Saxenian, *Silicon Chips and Spatial Structure: The Industrial Basis of Urbanization in Santa Clara County, California*. Working Paper no. 345 (Berkeley: Institute of Urban and Regional Planning, University of California, 1981).

34. Charles Piller, "The Cutting Edge: Focus on Technology; Innovation; Silicon Valley Immigration Debate Ignores Many Locals," *Los Angeles Times*, July 12, 1999, 1.

Chapter Five

Appropriating Technology

Alondra Nelson and Thuy Linh N. Tu interview Vivek Bald

Filmmaker Vivek Bald's most recent projects, *Taxi-vala/Auto-biography* and *Mutiny: Asians Storm British Music*, document, among other things, the use of technology by diasporic South Asian communities. At the center of these films are stories of innovations born out of necessity—the refashioning of everyday technologies for self-expression and political mobilization by immigrant communities. Here he discusses the "cauldron" of music, politics, and the DIY approach to technology.

Vivek Bald: To me music, technology, and politics are intertwined. And even if I wasn't able to articulate all the connections, I think I've been aware of them since I was really young. I grew up in Santa Cruz, California, which was an established stop on the Jamaican reggae tour circuit. So, from early on, I was exposed to all the greats of seventies reggae—Bob Marley, Peter Tosh, Lee Perry, Augustus Pablo, Big Youth, Black Uhuru, Hugh Mundell, Burning Spear. . . . That generation of Jamaican musicians were not only pioneers in music and their use of music technology but were also sending out a message of Third World liberation and self-determination. And as a teenager just starting to articulate a political consciousness, I found it was a message that really spoke to me, because it was echoed in my home.

When I was growing up, before we moved to Santa Cruz, my mom was teaching part-time at a University of California campus.

At that time, she was the only Indian woman teaching in her department and she became a kind of magnet for a lot of the graduate students who were from India, Pakistan, Afghanistan, Africa, Latin America, etc. So I had this strange situation: We lived in the middle of a white suburb, and for twelve years of my childhood we were never able to visit India, but our house would become a meeting ground for all these young leftist students from countries all over the world that had just gone through, or were still going through, independence struggles and various experiences of decolonization. It was in that kind of localized international community that I started to develop an understanding of what it meant to be of Indian descent, and an understanding of larger histories of colonialism and revolution. So when I got a little older and heard reggae music, it just brought all of that home.

A few years later, in Northern California, I became part of a punk scene, with anticorporate, antimaterialistic, anti-Reagan/Thatcher politics. Both of these musics—punk and reggae—were based on the appropriation of technology. Punk rock was driven by a DIY—"do it yourself"—attitude. If you had something to say, you just grabbed a guitar, a microphone, an amp—you started a band. You just got up there and said what you had to say. The same attitude applied to making 'zines—you just Xeroxed it and got it out there.

But this "DIY" approach to technology has always existed in poor communities, communities of color, Third World communities, out of sheer necessity. With limited resources and limited access, people have been using whatever technology they can get their hands on—secondhand, outdated, busted down, whatever—and have been pushing it, stretching it, redefining it, and usually getting it to do much more than it was ever meant to do. And that was the case with pioneer Jamaican producers like Lee "Scratch" Perry and King Tubby, who had music studios in Kingston filled with old, broken-down equipment, hooked and taped together, that they pushed in a way that redefined recorded music.

I think about that approach to technology too, when I think about places like the electronics market behind Jamma Masjid in Old Delhi, where you have men who sit in stalls surrounded by wires and transistors and soldering irons and bits and pieces of all kinds of equipment. I swear they can make anything out of anything. And I think you'd find electronics bazaars like these in just about any "Third World" or any formerly colonized country.

It was out of that cauldron of music, politics, and that DIY attitude toward technology and self-expression that I found a sense of identity and a place to articulate my politics. And that "place" wasn't specifically South Asian, but I don't think it really needed to be.

Taxi-vala

Vivek Bald: I moved to New York ten years ago to get a master's degree and explore the possibility of doing documentary film work. It was the first time that I had lived in a place with such a large and diverse South Asian community. What struck me right from the beginning was the stark class divisions. On the one hand, there were those in the community who had come to the U.S. between the 1950s and the 1970s who were mostly professionals—doctors, engineers, M.B.A.'s. And then on the other hand, there were more recent immigrants who were working at newsstands, restaurants, construction sites or were driving cabs. The first group, who had spent years courting and cultivating the whole model minority image, of course wanted nothing to do with the more recent immigrants. But I was much more interested in finding out what newer immigrants from South Asia were going through. When I found out that taxi driving—this quintessential New York job—was the most common job for new immigrants from South Asia, I decided to focus my first documentary on South Asian taxi drivers.

Thuy Linh N. Tu: And that was *Taxi-vala/Auto-biography*. What does "vala" mean?

Vivek Bald: "Vala" is a self-referent. It refers back to the word to which it is attached. For example, "the one who does this, or has this, or looks like this"; so taxi-vala, literally translated, is "the one who drives a taxi," or "taxi driver."

Alondra Nelson: Is there a reason that so many South Asians choose this industry in particular?

Vivek Bald: I think there are a number of reasons, but a lot of the drivers that I've talked to say it's because you are relatively autonomous. You set your own hours (within limits), you don't have a boss looking over your shoulder, if you save enough money, you can take a couple of months off and go back to your home country, then return to New York and start driving again right away. Which is not to say that drivers don't constantly have to answer to the police, the Taxi and Limousine Commission, and, of course, passengers. But it is also the quickest way to start earning some money. And you have the pattern where one person from a particular town or locality comes here and gets into a particular industry, and then helps others—friends, relatives, etc.—get into the same industry when they arrive. I think that's been a common pattern across all immigrant communities.

Thuy Linh N. Tu: How did you get started on this project?

Vivek Bald: I used the same DIY mentality that I got from punk rock; I started borrowing video cameras and going out and shooting. I used consumer Hi-8 cameras, a clip-on mic, and the help of one or two friends who would come out and shoot for me if they weren't busy on paid jobs. And if I couldn't get anybody, I just shot on my own as I interviewed. Hi-8 cameras were the best quality I could afford at the time. But documentary filmmaking isn't so much about high-end equipment as it is about content.

Alondra Nelson: One of the things that I noticed when I watched *Taxi-vala* was that taxi drivers still use CB radios. Can you talk about why this relatively low-tech device is still important to this community of drivers?

Vivek Bald: The first time I got into a cab in New York City, I noticed it had a CB radio, and I assumed that it was used for dispatch. Later I learned that yellow cab drivers were independent contractors and that there was no dispatch system. That's when I realized there was a whole other thing going on.

For anyone who grew up in the 1970s in the U.S., the CB radio has a very specific connotation. In the seventies the trucker—this burly redneck figure barreling down the interstates in a sixteen-wheeler with a CB radio in one hand—suddenly became an American pop cultural icon. There were trucker movies, trucker TV shows, "Keep on Truckin' " T-shirts, CB lingo dictionaries, people walking around saying things like "10-4 good buddy," and all of that. But in 1989 in New York City, CBs were being wielded in a whole other way, by a whole other group of people. I discovered pretty quickly that New York City taxi drivers were using different

CB channels to communicate with each other in different languages—Punjabi, Bengali, Greek, Russian, French Patois. Each driver community had their own channel. Here were immigrant workers who had appropriated this low-cost piece of equipment, formerly synonymous with redneck truckers, which they were now using to maintain unity and community in a job that by its nature separated them from each other and sent them off in different directions across the city.

And drivers use the CB radio at all levels. It's used for gossip, news, and work-related information. Drivers use CBs to share information about traffic pileups, police traps, and the best route to the airport at any given moment. Or they call each other for help if they are in an unsafe situation with a passenger, being harassed by the police, or TLC inspectors. What has developed over the CB airwaves is a set of "virtual communities" that predate the use of that term in relation to the Internet by five or ten years. And these really are "communities" much more so than the ones on the Internet, because the drivers who speak to each other over CB actually see each other. They eat together on their breaks, often live in the same neighborhoods in Brooklyn, Queens, or the Bronx, share the same histories, and share the daily experience as workers in a really exploitative industry.

Alondra Nelson: Is this virtual community open to all taxi drivers?

Vivek Bald: Well, yes and no. At some point when I was still working on *Taxi-vala,* I went down to Radio Shack and bought a forty-dollar CB radio. I rigged it so that I could plug it into a wall outlet and use it at home. Then I started tuning in to the stations that I had heard different drivers refer to, but I couldn't get anything but the occasional far-off trucker. So the next day I went in to the Lease Drivers Coalition office, where I was working, and told my friend Saleem what happened. He just smiled at me and said, "You don't have the crystal."

What he meant was that not only have drivers appropriated CB radios to their own uses, languages, etc., but they also figured out a way to place a chip or crystal in the circuitry of the CB so it can be tuned to the drivers' own private subfrequencies that lie in between the set channels of the radios. These subfrequencies can only be accessed by people who have "the crystal." One of the things that has spurred drivers to do this has been the instances of racist verbal abuse from non–South Asian drivers on open South Asian CB channels. The desire to create safe spaces to communicate and congregate is one of the things that has led them to adapt this technology.

Thuy Linh N. Tu: How similar is this South Asian immigrant CB culture to seventies CB culture? For example, do the drivers use pseudonyms or their real names? Is the dialogue formal or informal?

Vivek Bald: I think folks tend to use their own names, or at the very least nicknames that they might have within their immediate community or circle of friends. And it can be informal or more formal according to who is speaking to whom, and what they're talking about—just as in any community space. There are social structures within each CB channel that echo preexisting structures within the community. There are usually one or two people on any given channel who are more senior drivers—both in terms of their age and their experience in the industry—and are well-respected on that channel. Those drivers are often the most vocal, and the younger and more inexperienced drivers will go to them for advice about anything from how to get from point A to point B to how to deal with a ticket or summons.

Thuy Linh N. Tu: *Taxi-vala* ends with the striking image of cabs lined up along Broadway during a protest. Can you talk more about that protest and the role CB communities had in its organization?

Vivek Bald: You could see the political power of CB-based community maybe for the first time in October 1993, when there had been a series of driver murders—I think there were almost fifty murders that year. Most were of gypsy cab drivers; some were yellow cab drivers. In October, two Pakistani drivers were murdered in the space of just a couple of days. They were both drivers who were well known over the CB networks; one was a really young guy, in his early twenties, and the other was maybe the most well-known and well-respected voice on one of the major Punjabi CB channels.

After these murders there was a kind of spontaneous outpouring of outrage from the South Asian driver community over the city's indifference and inaction, and there was a massive demonstration, which came together almost exclusively over the CB airwaves. Three thousand drivers drove their cabs up the West Side Highway, across Forty-second Street, past the Taxi and Limousine Commission headquarters, then down Broadway to City Hall. By the time the first taxis reached City Hall, there was an unbroken line of cabs that stretched all the way back up Broadway to Forty-second Street. It was an amazing sight—this steady stream of yellow stretching up and down Broadway as far as the eye could see. The drivers couldn't drive any further, so they just left their cabs parked on Broadway and made their way on foot down to City Hall to demonstrate. This was the first massive show of strength and unity among New York taxi drivers in years, and it was the work of South Asian drivers and the networks and communities they had built using CB radios.

Alondra Nelson: Have there been any major changes in the driver community since you completed your documentary?

Vivek Bald: One thing that has started happening is that cell phones have now entered the picture. A lot of cell phone companies offer deals where you get nights and weekends free. There is a whole shift of drivers who work at night and sleep during the day, so most of the time that they are on their shift, they can have free calls. Some of the drivers who have been using CB radios in organizing efforts see that as a possible threat. With cell phones, interactions become very individualized, one person to another, in a way that undermines the more communal space of CB, where a number of people can listen and participate at the same time.

But cell phones have definitely not replaced CBs, and the people who are fighting the fight within the industry are making use of both technologies. The organization that I'm thinking of in particular is the New York Taxi Workers Alliance (TWA). From the start, TWA has been really strategic in the use of CB radio in their organizing efforts. They have worked closely with individuals who are prominent on their respective CB channels, who can get the word out about issues affecting the entire driver community, including demonstrations and ongoing campaigns. The TWA was the force behind an even more massive demonstration held in 1998—the stay-home strike that left New York streets almost completely empty of yellow cabs. That demonstration included a pointed use of CB and of channel "leaders" in order to involve as many ethnic and linguistic groups as possible. Whereas the 1993 demonstration was mostly South Asian, this time you had drivers from every single community participating. And that was in no small part due to strategic use of CB. But during the demonstration itself, cell phones were used to coordinate and communicate between people in different parts of the city, and to help coordinate press outreach. So there is a place for both technologies and the TWA have strategically used both.

Mutiny

Alondra Nelson: Your second documentary project, *Mutiny*, also looks at South Asian immigrant communities and their use of technology. But here you are concerned with the use of technology for creative expression. What was the genesis of this project?

Vivek Bald: Soon after I moved to New York, I started hearing this music that was coming over from Britain called *bhangra*. British bhangra was a reworking, or reinvention, of a Punjabi folk music that was originally tied to the harvest festival and is centered around the beat of the dhol, which is a large, two-headed drum. The sound of the dhol is loud and ferocious. In Britain in the 1970s, as the Punjabi community grew larger, more and more bhangra bands started to form to meet the increasing need for bands to play at weddings and other community functions. In the beginning these bhangra bands tended to be a mix of first-and second-generation immigrants. They started incorporating bits of "Western" technology—analog synthesizers, electric guitars, etc. And by the 1980s, you had second-generation British South Asian youth combining the beat of the dhol with hip hop, ragga, house, and techno. So a whole new bhangra was born, and along with it, a huge "underground" music industry that was largely cassette tape–based, and stretched from Britain back to the subcontinent, as well as to all points within the diaspora, including Toronto, Vancouver, and New York, and so on.

Initially, this was really exciting to me, because it was the first time that there was a music, or musical movement, that was specific to second-generation South Asians. But I got tired of bhangra relatively quickly when it became its own kind of mainstream within the South Asian community and got a bit glitzy and commercialized. (Now I feel somewhat differently—I don't think all bhangra is like that.) So I started looking for other second-generation British South Asian bands. I knew that the community in Britain was large enough and diverse enough that there had to be South Asian kids growing up there in the punk and reggae and hip hop scenes with the same musical-political sensibilities that I had.

And I discovered that there was, in fact, a whole other set of South Asian bands, musicians, and DJs working outside the bhangra scene. In a lot of cases, these were kids who were growing up in environments where South Asian, Afro-Caribbean, and working-class white youth were living together in a much more integrated way than you might see in the U.S., with a lot more cross-cultural exchange. The first group I started hearing about was Fun'Da'Mental, an agitprop political hip hop group that the British music press called "the Asian Public Enemy." Aki Nawaz, one of its founding members, was also an old punk—he was the original drummer for the band Southern Death Cult that later became the Cult. Around the same time, I also started hearing about groups like Cornershop, a guitar-based, grungy post-punk band using the occasional sitar or dhol or Punjabi lyric; the Kaliphz, who were a serious, straight-up hip hop crew from the northern industrial town of Manchester, with a mix of Asian, Black, and white members; and the Voodoo Queens, an all-female, mostly South Asian punk band.

Working on this project, I discovered that there was a whole other history of British Asian music. For example, South Asians were a major force in British hip hop; there were South Asian break-dance crews and South Asian graffiti artists all over the country. In interviews for *Mutiny*, the artists made frequent reference to Afrika Bambaata and Zulu Nation and to the Rock Steady Crew, who had come to Britain in 1981 and had a huge influence on the emergence of hip hop in the Asian community. During this time, there were also British Asian musicians using technologies like drum machines alongside traditional South Asian instruments. For example, at the age of fourteen, the producer and musician Talvin Singh was playing tablas alongside electro breakbeats for Asian break-dance crews. Similarly, brothers Farooq and Haroon Shamsher of the band Joi were mixing Bengali folk music with James Brown breaks.

Thuy Linh N. Tu: So you're suggesting that this music was born out of a hybrid urban culture.

Vivek Bald: Yes, but also, I think, out of the increasing racial intolerance in Britain during the time. After years of Thatcherite rule, Black and Asian youth in the late eighties and early nineties were growing up in the context of daily racial violence. And in a place like East London, where a lot of the Bangladeshi community was concentrated, you literally had generations of racists. In this part of London, there was a history of skinhead and white supremacist activity dating back to groups like the British Brothers League, through to the National Front in the 1970s, and the British National Party in the early 1990s. It was unsafe for Asian youth to be out on the streets past a certain hour, because there would be all sorts of drunken thugs piling out of the pubs looking for a "Paki" to beat up. And the police and local councils didn't do much to provide protection or make things any better. So growing up in that environment, Asian youth had to carve out spaces for themselves and build a certain level of strength and community and pride.

And music was central to that process. This music community was fostered in Asian youth clubs in East London where youth came together and put on raves and daytime parties. Musicians like Sam Zaman (State of Bengal), Deeder Zaman (from Asian Dub Foundation), Farooq and Haroon Shamsher (from Joi), and DJ/producers Ges-E and Ousman all have roots in Bengali youth organizations in East London. Probably the most famous was the League of Joi Bangla Youth, or Joi Bangla, which Sam, Deeder, Farooq, and Haroon were all involved in. Starting in the late 1980s, Joi Bangla created cultural awareness programs that brought together electro and hip hop music, break-dancing, and graffiti with Bengali dance, music, and culture. These programs were about getting Bengali youth to understand and express their position as people growing up in East London, at the same time as helping them learn about, and gain pride from, Bengali culture and history. It was a really formative moment for Asian youth in East London.

The group Asian Dub Foundation (ADF) grew out of a music technology workshop that was being run in East London for Bengali youth. ADF's bassist, Dr. Das, and guitarist, Chandrasonic, were both music technology tutors at a place called Community Music (CM), which is an independently run organization devoted to providing low-cost music training to people outside formal educational networks. The philosophy of Community Music goes beyond music training to developing the skills and self-confidence necessary to express what's inside oneself to other people. It's really about empowering people through music, with a focus on performance. As another member of ADF, John Pandit, puts it, "if you can get up and stand in front of people to perform music, you can get up and speak at a political rally."

ADF has set up their own organization back in East London called ADF Education, modeled after Community Music. There young people are being trained on a small, portable, and easy-to-use package of equipment that consists of a sampler/sequencer, a MIDI keyboard, a mixing board, and a mic. Because the equipment is so immediately accessible, the youth form a fast and close relationship to the technologies. Lisa Das, who runs ADFED, says that when many young people hit the age of about fourteen or so they stop being makers of music and become consumers of music instead. ADFED's aim is to catch young people at that moment and provide them with the skills and equipment to make music. ADFED gives them the tools to develop the inspiration, creativity, and talent that are already there.

Every step of the way ADF have epitomized what I talked about earlier as the intertwining of music, politics, and technology. Of the groups featured in *Mutiny*, ADF is now maybe the most successful group to have come out of the British Asian music scene. They are signed to a major corporate-owned label, London Records, and they're being really smart and strategic about being in that realm. They know that the label signed them because, at the moment, they are profitable, and a day will come when that is no longer the case and they will be dropped. So in the meantime they are using their position to spearhead campaigns like the one to free Satpal Ram, a British man of South Asian descent who has spent the last thirteen years in prison for defending himself against a racist attack. And they are funneling resources into ADFED.

People use the phrase "giving back to the community," but I

think ADF, Sam Zaman, and a number of others just have never left. They have their origins there and they are continuing the work, creating possibilities for demystifying and mastering technology for political and cultural self-expression.

Chapter Six

"Take a Little Trip with Me"

Lowriding and the Poetics of Scale

Ben Chappell

Introduction: Identifying Lowriders[1]

By most accounts, lowriding began in Los Angeles in the 1930s when some *mexicanos* gave their cars a low look by weighing them down, loading bags of cement or other heavy objects in the trunk.[2] The popularity of the low style grew in the 1940s and 1950s when the postwar economic boom afforded working-class *mexicanos*, many of them benefiting from veterans' pensions, unprecedented purchasing power at the same time that a glut of used cars flooded the market.[3] The classic lowriders are modified and decorated American-made sedans, but Japanese compacts, "Euros," trucks, and vans have also been turned into lowriders. In addition, a national lowrider bicycle trend is growing, with its own magazine and spaces reserved for customized bikes at any lowrider show. Some lowriders replaced their cars' suspension systems with hydraulic pumps removed from dump truck beds or military surplus airplane landing gear. The hydraulic suspension system was, by several accounts, a pragmatic modification that allowed cars to ride lower than the California legal limit, but then to be lifted in an encounter with a police officer.[4] The installation of hydraulics involves cutting the suspension springs to the desired (low) height, then installing lifts over each wheel, so the car's body can be raised while the wheel remains on the ground. These lifts are driven by a system of hydraulic pumps, in turn powered by an array of extra car batteries. The entire apparatus is installed in the trunk. The hydraulics are manipulated by means of a remote-control box (designed

Bike customization. Ben Chappell.

to match the decor of the interior), connected to the pumps by a cable, with switches for each lift so that individual wheels, an axle, or the whole car may be lifted at one time. The manipulation of "juice" or hydraulics has developed into a performative style that is unique to lowriders. On the street, hydraulics may be used to bounce the car, achieve "three-wheel" motion by tilting the car and leaving one tire off the ground, and "scrape" or "pancake" by raising and then dropping the car chassis to the street while moving, causing contact between the pavement and a metal "scrape plate" on the bottom of the car to throw a shower of sparks.[5]

In the 1960s the Chicano muralist movement contributed iconography and stylistic influence to the paintings that lowriders displayed on the bodies of their cars.[6] By the time the Chicano civil rights movement erupted in the late 1960s, lowriding was a well-established cultural practice in California cities.[7] In 1977 *Low Rider Magazine* (*LRM*) provided the lowriding community with a mass-media outlet,[8] and before 1980, lowriding had been "discovered" by mainstream publications including *Car and Driver, Natural History*, and the *New Yorker*. Lowriders also became important images in a burgeoning Chicano cinema, represented by such films as *Boulevard Nights*.[9] As the style gained notoriety, it was recognized in media discourse and in the limited scholarship on lowriding as an expression of Mexican American identity.

Brenda Jo Bright, the most accomplished ethnographer of lowriding, states that lowriding is concerned with "ethnicity as identity and ethnicity as style."[10] This description fits with one of the earliest "outsider" accounts, a 1976 article in *Car and Driver*, which noted the ethnic dimension of lowriding in no uncertain terms:

> 20 × 14 tires on tiny Cragers supporting a '64 Impala with no ground clearance [expresses] the refusal of young Chicano America to be Anglicized. There has never been a clearer case of the *automobile being used as an ethnic statement*. You can look at it from an automotive engineering standpoint and say it's an atrocity. But if you do, you haven't seen it. This isn't engineering, this is community consciousness.[11]

This sentiment is echoed in the statements of lowriders themselves, who often reiterate the popular belief that "[t]his is ours. This is what we started. This is what we do."[12] As the public relations officer of a Latino fraternity told me at a car show on a Texas university campus, "That's what it's about: expressing our culture."

If lowriders are an expression of identity, what specifically is this lowrider identity? Several writers locate the ethnicity of lowriders in aesthetic and performative links with Mexico, identifying what they apparently interpret as residual Mexican "survivals" in the repertoire of lowrider signs. The southwestern folklorist James Griffith found a baroque sensibility in interior customization among Arizona lowriders: he reads "the opulence of materials and appointments, . . . their interest in miniaturization, and especially . . . their concentration and accumulation of independently significant details" as a reference to the Spanish missions and eighteenth-century chapels of the southwestern United States, and thus to a historical Mexican-ness, recalling that much of the region was in fact part of Mexico until 1848.[13] While this explanation is convincing in the case of lowriders that draw on specifically Mexican-national or Mexican Catholic imagery, the idea of lowrider identity as indexical to Mexico does not hold with lowriding in general: lowriders seen on the street or at shows in Texas or featured in the pages of *Lowrider Magazine* often draw on images from comics, rap music, or other sets of signs from the urban United States and its media culture. In light of these images, to argue that lowriders identify principally with a Mexican national identity seems reductive; rather, lowriders represent a hybrid "urban folk culture" in process.[14]

The folklorist William Gradante located lowriding within a larger history of communally made public art projects, namely the muralists movement in revolutionary Mexico early in the twentieth century.[15] Although this position again risks reductive national identification, it does describe attitudes evident in the collaboration among fellow car club members that often goes into building a car. For example, the member who is handy with an airbrush might paint murals for all fellow members, and one who works at a body shop might help clubmates with more radical modifications.

This accounts for the connection between lowriders and the decidedly non-Mexican "kustom kar" movement of the 1950s. Several accounts acknowledge Harry Westergard, a non-Chicano in 1930s Sacramento, as the first to lower a car for aesthetic reasons by mechanically modifying it, as opposed to loading it with extra weight. Westergard taught his auto-mechanical modification techniques to youths in his neighborhood. One of these protégés was George Barris, who brought his craft to southern California and made a name for himself as the "King of Kustomizers" in the 1940s and 1950s. He eventually designed cars such as the Batmobile for Hollywood and influenced not only major automotive designers but perhaps the first *mexicano* lowriders with his baroque modification of automobile bodies and brilliant finishes.[16] This history has lowriders emerging as "part of a broad interest among working-class youth in making cars unique in some way."[17] Placing a high social value on automotive mechanical skills can be related to a working-class need to extend the functional life of a car and to modify its appearance when the economic means to buy high-status vehicles are limited.[18] Among the modifications typical in this practice are "chopping," or lowering the roof of the car by cutting and rewelding it with a torch; "frenching," or recessing antennas or headlights into the body of the car; and installing "suicide doors," which open at the front by means of hidden switches and are hinged at the rear, all of which are undertaken not only by lowriders but by other (often non-Chicano) car customizers.[19] This history suggests a view of lowriding as a subset of the broader American car culture.[20]

Some situate lowriders within a genre known as the "lead sled," a smooth, sleek car modification style that follows a "basic American aesthetic preference" in its lines.[21] The smooth, clean lead sled style may resemble some lowriders on the exterior, but the baroque decorative lowrider interiors offend the lead sled sensibility. This "dressed-up" look was once rejected vehemently enough by lead sledders to inspire a racial slur: a car with superfluous decoration was considered "niggered up."[22] One of the distinctive characteristics of lowrider style is in fact the way lowriders allow smooth lines and luxurious frills to coexist.

But a lowrider is defined not only by how it looks, but also by how it is made and displayed. Thus lowriding is a performance genre in the most literal sense, and identification happens in the *process* of making and driving a lowrider. Lowrider performance is also motivated by different mechanical/kinesthetic ends than broader American car culture. While (largely Anglo) customization in the hot rod genre is geared toward efficiency and "performance," seeking to maximize the potential for speed,[23] lowrider modifications have adverse effects on typical automotive performance since lowrider cruising emphasizes style and socializing, not speed:

> Slow driving helps human eye contact and creates opportunities for favorable responses from onlookers (eye movements, smiles, words, gestures). Social interaction becomes the raison d'être for driving, and why move quickly when it is social contacts one desires? Use of the car in this fashion increases territorial coverage and thus opportunities to meet more people, particularly members of the opposite sex.[24]

Hence the alternative performance standards expressed in the lowrider manifesto: "low and slow."[25]

Lowriders are thus marked as belonging to a distinct genre of expression in the ways they differ from mainstream and other subcultural notions of what a car should do. They are about form, not function; sociability, not speed. By driving a "Mexican" car, lowriders distinguish themselves in traffic first from the middle-class majority, driving their standard-issue cars, and from customizers who modify for speed and efficiency, such as hot-rodders.

The social networks and mutual influences of car clubs, lowriding publications, and contests and reactions to the expressive patterns of other groups—such as the Anglo California rake—shape the distinctions of lowrider culture. Lowriders move in and out of Anglocentrically acceptable social practice, between positions within mainstream consumption patterns of commercial culture and those outside them.[26] In fact, lowriding reflects both the growing *mexicano* middle class in the post–civil rights movement era and aspects of class aspiration among a more marginal urban population, emerging as what the esteemed anthropologist Renato Rosaldo calls a new form of "polyglot cultural creativity."[27]

It is important to note that lowriders are a part of a community with a long memory of *mestizaje*. The Mexican *raza mestiza* is a "cosmic race," that represents the offspring of the Spanish conquerors and the Native Americans whom they found here.[28] The condition of *mestizaje* thus problematizes binary oppositions: a *mestiza* person is descended from both the conqueror and the conquered, but is neither. The *mestizaje* concept is a narrative version of the hybridization that, through a combination of the miniature and the gigantic, produces a polysemic yet grounded lowrider style that fits with Mexican American experience of hybridity.

The Miniature and the Gigantic

Some of the most noticeably distinctive characteristics of lowriders involve parts that are smaller, lower, or more elaborate than those on other cars. Because so much of the lowrider aesthetic involves playing with conventions of scale or degree and the setting and breaking of limits, Susan Stewart's theorization of the poetics of scale—the miniature and the gigantic—is helpful.[29] Miniaturization has been noted as a key aspect of lowrider style.[30] "Undersized" wheels and tires are perhaps the most immediately visible features of a lowrider. The accessories that decorate the interior of a lowrider are often small versions of luxury items—chandeliers, televisions, bars, or fountains. Griffith's thesis that such baroque miniaturization references Catholic cathedrals is tempting, but becomes somewhat tenuous when so many miniatures in lowrider interiors are models of modern luxury items. Stewart's reading of the miniature is more compelling: like an adult's dollhouse, a quintessential expression of the miniature, a customized lowrider interior "has two dominant motifs: wealth and nostalgia. It presents myriad perfect objects that are, as signifiers, often affordable, whereas the signified is not. Use value is transformed into display value here."[31] This transformation is also reenacted in building minia-

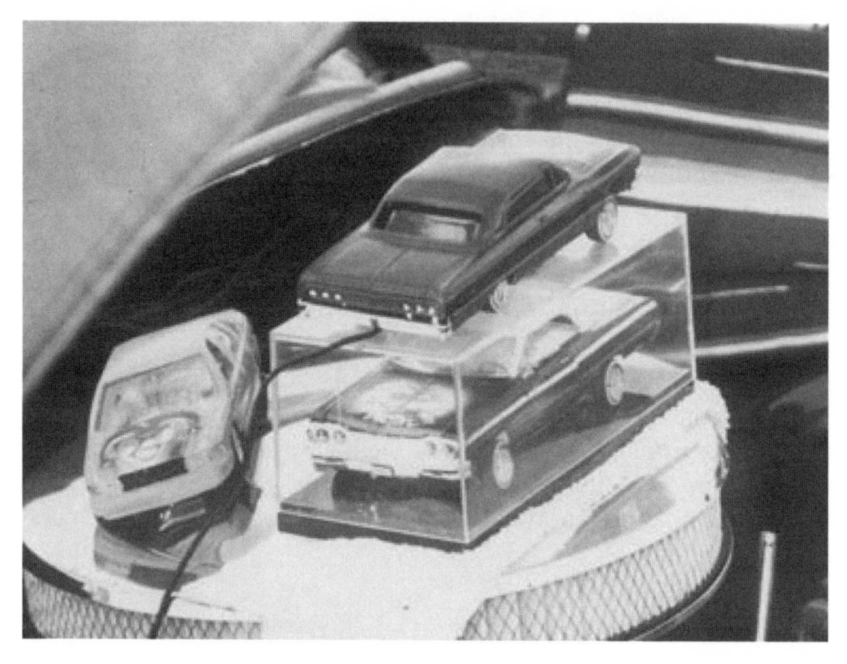

Lowrider miniatures. Ben Chappell.

ture model lowriders, which is a customization genre in its own right, often attracting those who are too young to drive or who cannot afford their own cars.

The miniature for Stewart represents nostalgia for preindustrial production, a preference for the kind of artisanal workmanship that a lowrider clearly displays. Initially products of assembly lines, lowriders are made unique by their owners. Lowriders on display at a show are surrounded by trophies from prior shows or magazine issues that have featured them, framing the cars as valued, singular works of art. Every display car is also given a name, and individual cars acquire fame by being shown and winning awards on the car show circuit.[32] In a lowrider, then, working-class mechanical skills are translated into the creation of individuated, aestheticized display objects that index the spending power and leisure time necessary to create them. For Mexican Americans, this can be seen partly as an act of resistance against the ascription of a "commodity identity" that labels them as "cheap labor."[33] This works through both attention to luxury, opulence, and status objects and a celebration of the mechanical skills required to modify a car. A bona fide lowrider should be one-of-a-kind, and not purchased already fully customized.

While a "do-it-yourself" ethic and resistance against the deterioration and devaluation of working-class mechanical autonomy is one face of lowriding, another side of the aestheticization of a car is the expression of class aspiration. As J. T. Borhek, Michael Stone, and other writers point out, "class" has been a positive term in lowriding for decades, and is often used synonymously with "pride." In this light, small parts and baroque luxury on a lowrider reflect the quest for control, completeness, and perfection articulated as the project of miniaturization by Stewart. The regal atmosphere of a lowrider interior creates a tiny, rolling dream mansion. This contributes to an interiorization of the car—it becomes a movable home for people whose mobility is sharply limited by police surveillance and harassment as a result of racism, and a personal sanctuary for young working-class people who might share limited domestic space with large extended families.

Still, the overdecoration of a mundane object (the car) offends bourgeois ideas of "responsible" accumulation and consumption in a way that touches on race, class, and urban markers much like the dandyism of a 1940s zoot-suiter (see discussion below) or contemporary rap music stars. The opulence of lowrider interiors and their elaborate paint jobs or body modifications take consumption to carnivalesque extremes. In this surpassing of Anglocentric limits, lowrider style resonates with what Stewart terms the gigantic: "In contrast to the still and perfect universe of the miniature, the gigantic represents the order and disorder of historical forces."[34] The gigantic is the scale of lurching fluidity, of boundaries broken, of the excessive. Similar to Bakhtin's concept of the carnivalesque in its grotesque transgression of the bounds of "good taste," the gigantic represents the popular, the mass, the "anti-cultural."[35]

A gigantic disruption of propriety can be found in the economics of lowrider style. Building a show car can easily cost over ten thousand dollars, and the best-known celebrity cars require $200,000 or more.[36] Working-class lowriders are forced to develop their cars gradually, often showing them as they progress, adding a new feature for each show in hopes of garnering a few more points from show judges, and bartering with family, friends, and acquaintances when cash is in short supply.[37] The hydraulic system is often the biggest expense, and the chrome or gold plating of engine and body parts, structural modifications that require welding, and paint jobs further pad the bill. This is the dollhouse made gigantic—display value out of place. It is on the street rather than in the home; it is an overdecorated used car rather than a more "appropriate" status symbol or "responsible" use of financial resources.

Such acts of conspicuous, challenging consumption have a long history in minority urban popular expression, going back at least as far as the *pachucos*.[38] Dressed in their flamboyant zoot suits or "drapes," accessorizing with wide-brimmed hats and extremely long watch chains, speaking in a syncretic mixture of *caló*[39] and hip, jazz-cat slang, Los Angeles *pachucos* in the 1940s practiced a highly visible form of youth culture.[40] Although the zoot suit was popular around the country, among diverse groups, it became particularly associated in Los Angeles with the *mexicanos*, the largest, most visible minority group in the city at the time.[41] The zoot-suiter *pachucos* occupy a special place in the collective memory of lowrider culture. Although their legendary status as the first Chicanos to lower their cars has been challenged as ahistorical, there is no doubt that the image of the *pachuco* is salient in current lowrider practice.[42] *Lowrider Magazine*'s logo is a stylized *pachuco* face, nostalgic images of zoot-suiters abound in lowrider mural art, and zoot suit modeling contests have been a staple of lowrider shows.[43]

Paralleling the gigantic performance of consumption in lowrider style is the gigantic movement enabled by hydraulics, especially in the heightened aesthetic context of a car show. The manipulation of hydraulic systems as a form of dance or sport is a decidedly inappropriate use of the automobile by the standards of the automotive industry (although a whole alternative industry has grown up around manufacturing hydraulic equipment for lowriders). When lowriders hop their cars, they engage the hydraulic lifts to raise the car, then let it drop, engaging the lifts again just as the suspension springs recoil. In this way, they can gradually increase the height of the bounce, until one axle, or sometimes the entire car, hops up to five or more feet off the ground. The image of a flying, bouncing car is comical and wild; it is a gigantic bursting of limits imposed by gravity and propriety.

Tendencies toward the gigantic are played out in the other major genre of lowrider car show performance: hydraulic car hopping and dancing. In contrast with the display of a show car, car hopping disrupts the tightness of a closed boundary and flirts with a loss of the control that is so complete in the exhibition display. Although car hopping developed in informal gatherings of lowriders in parking lots and other public places, the rise of lowrider car shows or "happenings," largely under the auspices of *LRM*, has led to the codification of competitive hopping events. At some shows a purse of several hundred dollars goes to the winner, allowing a semiprofessional star system to develop among hoppers. The regular car hop is the most basic hydraulic competition. The goal is to bounce one axle for height, which several judges determine by holding a large ruler close to the bouncing wheels. Standing outside the car, the contestant holds the control box, with the cable running into the car's open window. Engaging the hydraulics rhythmically, the contestant causes the car's front wheels to bounce off the ground, and the prize goes to the car with the most "bounce to the ounce," or sheer height off the ground.

A more freestyle form of hopping competition is the car dance. Again standing outside the car, manipulating the controls much like the joysticks or buttons of a home video game unit, the contestant puts the car through a repertoire of movements, including bouncing one axle, back or front; bouncing both axles until the car jumps and all tires leave the ground; rocking the car from one side to the other; and wiggling back and forth in small, fast movements between the two tires on one axle. These manipulations of machinery visually resemble gigantic versions of other technological pastimes like playing video games or driving miniature remote-control model cars. In lowrider car hopping, the object of remote control is made material and huge.

If lowrider style is a certain kind of Mexican American culture in action, it "prioritizes" an urban Mexican American identity.[44] Does understanding "culture" as processual and lowriders as hybrid culture mean that "anything goes" in lowrider style?[45] Like any expressive genre, lowriding has its rules. We can further illuminate, the dynamics of identification around lowrider practice by looking at ongoing contests within the lowrider world: different groups who associate with the style have different ideas about what lowriders mean.

Urban *mexicano* street style is closely associated in the U.S. popular imagination with social organizations considered to be "gangs."[46] Some *mexicano* street gangs in the United States are in their third generation, dating back to the 1920s, and expressive forms like customizing low cars and writing stylized graffiti have long been important activities of these groups.[47] The institution of lowrider car clubs, however, was a project of the Los Angeles County Sheriff's Department and a group of merchants as an anti-gang measure:

> Originally designed to provide an alternative to gangs, car clubs became a focal point for social life in the *varrio*, providing a place to work, hang out, listen to music, gain knowledge of self-expression and cultural identity through the art of car customizing.[48]

Since the boom in the popularity of lowriding in the 1970s and early 1980s, a sort of "cultural ambassador" discourse has emerged among car clubs and in *LRM* and other observers' accounts as a reaction to the associations with "gang culture." Car club members in particular are represented as trying to create a positive image of Mexican Americans: "the low rider enthusiastically shares his cultural heritage and mechanical and artistic creativity with the American majority culture. He does this without abandoning the values of his community."[49] Despite such diplomatic attempts, lowriders have met with repression from authority figures located both within and outside their communities. At an Austin lowrider show, I spoke with a member of the Latino fraternity at a local university. Francisco said he likes lowriders and going to car shows, but his parents didn't want him to get involved with lowriding when he was growing up in Dallas because "People see it as a gang thing."[50] He knew people who had been pulled over by the police for "no probable cause, no speeding, just because they've got rims." Enrique, an older *veterano* who claims to have first brought lowrider style to Austin from California in the early 1970s, also speaks of antagonism from the police, telling the story of being pulled over on the first day he drove his lowrider in town because the officer "didn't like the looks" of his low car. Such stories abound in lowrider circles.

The conflict between lowriders and police is symptomatic of larger cultural patterns that criminalize young urban minority males. In a study of East Los Angeles Chicano gang members, the sociologist Joan Moore observed that the stigma associated with nonwhite racial identity causes "ascribed deviance."[51] Part of this ascribed deviance is presumed guilt by association—since some gang members are interested in lowriders, lowriders are assumed to be gang members.[52] Lowriders and Chicano gang members do share a certain cultural vocabulary and set of historical referents—the zoot-suiter, terms like *la vida loca* (the crazy life)—and other aspects of the urban *barrio* experience: stigmatized with a racial identity interpreted in the public culture as criminal, the person of color is subject to limited mobility, high levels of surveillance, and curtailed economic and educational opportunity in de facto segregated cities under a "border-patrolled state."[53]

Members of lowrider car clubs respond to this labeling by constantly asserting that their art is "not a gang thing." Sometimes this is the organizing principle behind what I call lowrider club activism. An Austin lowrider activist, Miguel started his own lowrider car club when he felt that the club he was involved with was drifting toward gang activities.[54] In describing this, Miguel articulates a theme that has been part of participants' discourse on lowriding since the style was first documented—an ongoing tension between "gang culture" and "positive" alternatives. Both sides in this contest claim an authentic relationship with expressive street culture, but activist lowriders proclaim their clubs to be antithetical to gang involvement, which they see as characterized by drug use and trafficking, truancy, and violence.

Lowrider activists use expressive cultural activities as a social intervention, particularly on behalf of young teenagers and preteens, who may be facing the pressure of gang recruitment for the first time. This is the motive behind Austin lowriding youth groups such as Creative Touch, a bike and model car club, and Mystic Knights, a bike club. Creative Touch's president, José, is a single parent of two teenagers. José used to have a lowrider car, but "when I got custody of the kids, I couldn't afford both, so I got rid of the car." José learned the art of customizing model cars from his father, and rigs his models with miniature, battery-powered lift systems like the hydraulics on life-sized lowriders, model wire-spoke wheels, and the same gleaming, multi-coat paint jobs as on the real thing. When I met him, José emphasized to me that his club was an alternative to gangs for the kids involved. He expressed frustration with parents in the community who identify lowriding with gangs. Numerous parents share the fears of Francisco's, and will not allow their kids to participate in his club.

Members of Miguel's club spoke to me about ex-members who had been unable to meet Miguel's strict requirements for membership in his group and had been excluded or had quit. This is one of the direct forms that the activist clubs' interventions take: in order to remain part of the club, members must meet requirements that often include, as in the case of Miguel's club, being "drug-free" and "gang-free," participating in community service projects, and staying in school with passing grades. Since lowriding is viewed by youths as an authentic identity practice (that is, as performing Chicano identity), positive clubs like José's and Miguel's are attractive to young, urban *mexicanos*; membership is a reward granted to those who demonstrate the desired behavior. Representatives of activist clubs speak of their social intervention as though it were in itself the most authentic activity of lowriders, in contrast with the actions of gang-associated lowriders who have strayed from the "correct" path. Bright relates similar stories of lowriders who felt that participating in a positive car club and building a car constituted one of the few social opportunities available to them outside *la vida loca*, or gang life on the street.[55] Thus lowrider style, as part of the fabric of urban *mexicano* communities in the United States, is a resource that can be tapped into by activists.

Conclusion

Of course, it is difficult to speak about a unified "Mexican American identity," especially with reference to such a gendered activity as lowriding.[56] Lowriding is influenced by the history of automotive mechanics, and the "street" in general, and the patriarchal (not only Mexican American) assumptions that both of these are the domains of males. Women are not absent from lowriding, but their presence is often limited to swimsuit modeling.

In 1979, *LRM* first featured a woman in a swimsuit on its cover, and was soundly criticized by both male and female readers. The strongest opposition seemed to come from those who saw this as a "trashy" representation of a "nice homegirl."[57] The editors pointed to the 15–20 percent leap in circulation that followed, however, and on this basis have rationalized making similar images defining feature of every issue since. Other publications, such as *Orlie's Lowriding, Street Customs*, and San Antonio's *Vajito*, also prominently feature glamorous images.[58] As a national circuit of lowrider car shows developed, "bikini contests" and "bad girl contests" became predictable parts of the program, in which female audience members are invited to participate in a sort of beauty pageant.[59]

In 1996 the Xicana Caucus of Movimiento Estudiantil Chicano/a de Aztlán (MEChA) introduced a resolution calling for a boycott of *LRM*, in protest of the objectification of women on its cover. The caucus's critique resonated to a subordinate position in the Mexican American community.[60] The resolution did not pass a general vote but it prompted *LRM* to respond. *LRM* in turn accused the women of being divisive, of betraying *LRM* after the magazine had supported MEChA and other movement causes.[61] The publisher of a Texas lowriding magazine expressed a similar position to me, claiming that he does models a favor by giving them career opportunities that they might not get from mainstream publications. Further, in a letter to MEChA, women employees of *LRM* rejected what they interpreted as a puritanical attempt to censor women from displaying their bodies without shame.[62]

It is important to note that criticism such as that of the Xicana Caucus does not deny the Chicano identification with lowriders in general, though they take issue with a particular aspect of how lowrider culture presents itself. The idea that lowriding can be salvaged from stifling patriarchal practice lends power to the implicit criticism by female lowriders that lowrider culture is guilty of gender exclusion and objectification of women. This idea also underlines the depth of the identifying links between lowriding and Chicanos. There are women who build their own lowriders, and even command professional fees for displaying them as featured cars at shows. Some women pose in front of their own cars in swimsuits for *LRM* photo sessions. Others eliminate the ornamental model role altogether and pose in street clothes: the mechanic/owner and her car. These are all examples of how gender roles in lowriding are not static, but under negotiation. At least two all-female lowrider clubs are active in San Antonio, and present a plausible, if optimistic, vision of what lowriding without patriarchy might someday look like.[63]

One of the most optimistic lowriders I have yet seen is a six-year-old from San Antonio who showed her lowrider tricycle at an Austin bike show. Called "Lokita's Trike," this little machine was gigantically accessorized, decorated with a velvet seat, a trailer for her teddy bear, and plenty of chrome trim. Lokita's parents both helped her carry a trophy bigger than she was away from the judges' table at the end of the day. Lokita is growing up in a very different era than prior generations, one in which feminism has begun to denaturalize gender inequity through debate, and in which "positive" lowriding is taken for granted by participants as part of lowrider practice. Her parents, by involving her in this activity, are providing her with successful experiences and expressive tools not unlike those that some children garner from little league or music lessons. When Lokita is old enough to drive a car, lowriding will no doubt look very different than it does now, but it can be expected to continue to serve as a site of identification and negotiation for its participants.

NOTES

Thanks are due to Deborah Kapchan and Mieke M. F. Curtis-Richardson for their thoughtful critiques of a draft of this chapter and to Ma. Angélica Bautista Farías for sharing her excellent and instructive M. A. report with me at a key point in the writing process. I am also indebted to Paul Erickson for making me aware of this collection. As always, the most indispensable support and the first and last lines of editing advice came from Marike Sophie Janzen.

1. Following my sources, I use "lowrider" to denote both the car and the human participant in lowrider culture. Context will make clear which meaning is in effect. In addition to the works cited in the notes, this chapter draws on field research at several lowrider events in Texas between the spring of 1997 and the winter of 1998. Since I have not undertaken fieldwork for an extended, continuous period of time, however, I offer the present arguments as hypothetical, not ethnographic. By the term "lowrider community" I refer simply to those who participate in lowrider style as car owners or builders, club members, attendants at car shows, or readers of lowrider publications.

2. See Brenda Jo Bright. "Mexican American Low Riders: An Anthropological Approach to Popular Culture" (Ph.D. diss., Rice University, 1994); and Calvin Trillin, "Our Far-Flung Correspondents: Low and Slow, Mean, and Clean," *New Yorker*, July 10, 1978, 70–74.

3. Lowriders appropriated techniques of body modification and finishing from an automobile customization subculture that had also developed in the 1950s. Michael C. Stone, " 'Bajito y Suavecito': Low Riding and the 'Class' of Class," *Studies in Latin American Popular Culture* 9 (1990): 85–126. See also Tom Wolfe, *The Kandy-Kolored Tangerine-Flake Streamline Baby* (New York: Pocket Books, 1965).

4. Wayne King, "Low Riders Are Becoming Legion among Chicanos," *New York Times*, May 9, 1981, 8; Trillin, "Our Far-Flung Correspondents"; Ted West, "Scenes from a Revolution: Low and Slow," *Car and Driver*, August 1976.

5. Trillin, "Our Far-Flung Correspondents."

6. For further discussion, see Sojin Kim, *Chicano Graffiti and Murals* (Jackson: University of Mississippi Press, 1995); and Ricardo Romo, "Borderland Murals: Chicano Artifacts in Transition," *Aztlán: A Journal of Chicano Studies* 21, nos. 1–2 (1992–96): 125–54.

7. Paige R. Penland, "The History of *Lowrider Magazine*: What a Long, Low Trip It's Been," *Lowrider Magazine* 19, no. 1 (January 1997): 74–82. *Con Safos* later served as a model for the publishers of *LRM*.

8. The magazine's current incarnation is entitled *Lowrider Magazine*, the change in title coming when the publication folded for financial reasons in 1985 and was relaunched in 1987. Started with student funds by Chicano student council members at San Jose State University in California, *LRM* grew out of the involvement of college students in the Chicano movement to become, by its own estimation, "the most successful Chicano publication ever." See Penland, "The History of *Lowrider Magazine*." Currently distributed around the United States and in several other countries, *LRM* has grounds for this claim.

9. Dealing with "cruising" culture and street gangs, *Boulevard Nights* is an important historical example of the conflation of Chicano street style and social deviance. See the discussion of lowriders and gangs below.

10. Bright, "Mexican American Low Riders," 33.

11. West, "Scenes from a Revolution," 76.

12. Bright, "Mexican American Low Riders," 37.

13. James Griffith, "Mexican American Folk Art," in *From the Inside Out* (San Francisco: Mexican Museum, 1989), 55, 58. Also see Rodolfo Acuña, *Occupied America: A History of Chicanos*, 2d ed. (New York: Harper and Row, 1981).

14. Brenda Jo Bright, "Remappings: Los Angeles Low Riders," in *Looking High and Low: Art and Cultural Identity*, ed. Brenda Jo Bright and Liza Bakewell (Tucson: University of Arizona Press, 1995).

15. William Gradante, "Art among the Low Riders," in *Texas Folk Art*, ed. F. E. Abernathy (Dallas: Southern Methodist University Press, 1985).

16. Stone, " 'Bajito y Suavecito,' " 97–98; and Wolfe, *The Kandy-Kolored Tangerine-Flake Streamline Baby*.

17. Luis F. B Plascencia, "Low Riding in the Southwest: Cultural Symbols in the Mexican Community," in *History, Culture and Society: Chicano Studies in the 1980s* (Ypsilanti, MI: Bilingual Press/National Association for Chicano Studies, 1983), 144.

18. Evan Watkins, " 'For the Time Being, Forever': Social Position and the Art of Automobile Maintenance," *Boundary 2* 18, no. 2 (1991): 150–65.

19. Borhek, "Rods, Choppers, and Restorations: The Modification and Recreation of Production Motor Vehicles in America," *Journal of Popular Culture* 22, no. 4 (1989).

20. Stone, " 'Bajito y Suavecito,' " 95.

21. Moira Harris, *Art on the Road: Painted Vehicles of the Americas* (St. Paul, MN: Pogo Press, 1988), 69.

22. Borhek, "Rods, Choppers, and Restorations," 101.

23. Borhek, "Rods, Choppers, and Restorations," 100.

24. James Diego Vigil, *Barrio Gangs: Street Life and Identity in Southern California* (Austin: University of Texas Press, 1988), 122.

25. William Gradante, "Low and Slow, Mean and Clean," *Natural History* 91, no. 4 (1982: 29–39; Trillin, "Our Far-Flung Correspondents."

26. Bright, "Remappings: Los Angeles Low Riders," 97.

27. Renato Rosaldo, *Culture and Truth: The Remaking of Social Analysis* (1989; Boston: Beacon Press, 1993), 215.

28. Gloria Anzaldúa, *Borderlands/La Frontera: The New Mestiza* (San Francisco: Aunt Lute Books, 1987), 77 (emphasis in original).

29. Susan Stewart, *On Longing: Narratives of the Miniature, the Gigantic, the Souvenir, the Collection* (Durham: Duke University Press, 1993). Looking at literature, folklore, and leisure activities, Stewart describes how everyday "narratives" of the gigantic and the miniature have functioned in Western, bourgeois expression. According to Stewart, the miniature as an aesthetic value expresses bourgeois longing for control and secured boundaries through a heightened sense of interiority and escape from the real world in objects such as toys, models, miniature furniture, and the dollhouse. The miniature emerges as a trope in Western culture through the leisure activities of adults around such objects. The miniature is

> nostalgic in a fundamental sense. . . . [Miniature objects] completely transform the mode of production of the original as they miniaturize it: they produce a representation of a product of alienated labor, a representation which itself is constructed by artisanal labor. . . . Whereas industrial labor is marked by the prevalence of repetition over skill and part over whole, the miniature object represents an antithetical mode of production: production by the hand, a production that is unique and authentic. (58, 68).

This backwards glance to preindustrial handicraft performs a white-collar distance from industrial production, a privileged position in modernity.

30. For example, Griffith, "Mexican American Folk Art."

31. Stewart, *On Longing*, 61–62.

32. "Wild Thing," "Wicked Six," "Blue Bird," "Keep It Live," and "Below Average" are only a few examples from a diverse pool of names.

33. Carlos G. Velez-Ibañez, *Border Visions: Mexican Cultures of the Southwest United States* (Tucson: University of Arizona Press, 1997).

34. Stewart, *On Longing*, 86.

35. Mikhail Bakhtin, *Rabelais and His World*, trans. Hélène Iswolsky (Bloomington: Indiana University Press, 1984).

36. Celine García, "Putting Dreams in Gear," *Austin American-Statesman*, October 14, 1995, B5.

37. Ferreira, *Blue Magic Can Happen* (Hayward, CA: Vida Vision, 1990).

38. According to Ruben Martínez, the *pachucos* were "the first style-conscious Latino gangs." *The Other Side: Fault Lines, Guerilla Saints, and the True Heart of Rock 'n' Roll* (London: Verso, 1992), 109. See also Maricio Mazón, *The Zoot-Suit Riots: The Psychology of Symbolic Annihilation* (Austin: University of Texas Press, 1984).

39. Often misrepresented as a mixture of Spanish and English, caló is a derivative of zincaló, an amalgam of European and Semitic languages tracing back to fifteenth-century "Spanish gypsies" and brought to the New World by members of the conquistador parties or perhaps by colonial bullfighters. Mazón, *The Zoot-Suit Riots*. The North American version of caló, which adds English to the mix, was spoken by the *pachucos* of Los Angeles and the Southwest, and is still popular among *cholos*, the contemporary incarnation of the *pachucos*, and in the pages of lowrider publications.

40. James Diego Vigil, "Car Charros: Cruising and Lowriding in the Barrios of East Los Angeles," *Latino Studies Journal* 2, no. 2 (1991): 74.

41. Acuña, *Occupied America*, 324.

42. Plascencia, "Low Riding in the Southwest."

43. Bright, "Mexican-American Low Riders."

44. I borrow this phrase from Tricia Rose's exemplary work on what I consider another urban folk culture: "Rap music is a black cultural expression that *prioritizes* black voices from the margins of urban America." *Black Noise: Rap Music and Black Culture in Contemporary America* (Hanover: Wesleyan University Press, 1994), 2 (emphasis added). Lowriding and hip hop share not only parallels in aesthetics and poetics, but often are performed together. See Ben Chappell, "Genre Affinities in Urban Expressive Culture: Prospects for Ethnography of Lowriders and Hip Hop" (paper presented at the American Folklore Society annual meeting, Austin, Texas, October 29, 1997). Lowrider style has provided key signs for the album covers and music videos of Los Angeles rappers such as Ice Cube and Kid Frost. Also, lowrider shows often feature performances by rappers from *LRM*'s subsidiary record label Thump Records and others. See Steve Loza, "Identity, Nationalism, and Aesthetics

among Chicano/*Mexicano* Musicians in Los Angeles," in *Selected Reports in Ethnomusicology: Musical Aesthetics and Multiculturalism in Los Angeles,* ed. Steve Loza (Los Angeles: Department of Ethnomusicology and Systematic Musicology, UCLA, 1994), 54.

45. Juan Gómez-Quiñones, *On Culture;* Popular Series no. 1 (Los Angeles: UCLA-Chicano Studies Center Publications, 1977); Rosaldo, *Culture and Truth.*

46. The term "gang" carries a strong political charge. In cases of police harassment of lowriders, it would sometimes seem that a gang can be defined as four minority men in the same car. Machine Gun, a rapper with the politically left hip hop group Aztlán Nation, engages this definition in his "pro-gang" stance:

> just because the media or whoever [sees] a group of Chicanos and they say "Oh, that's a gang" and they get scared. I grew up in the barrio and we had a gang, a clique; we took care of each other. It was like a second family. The thing that I'm against is gang-banging, I'm against Raza killing Raza for bullshit territory.

In Arnoldo García, "Aztlán Nation Rapping por la Raza," *Crossroads* 22 (June 1992): 24–25.

In contrast with Machine Gun's recuperation of "gang" and the Spanish *clica,* most activist lowriders I have spoken with use "gang" pejoratively to refer to a range of behaviors viewed as negative: predominantly violence and drug use.

47. Vigil, *Barrio Gangs;* Marcos Sanchez-Tranquilino, "Space, Power, and Youth Culture: Mexican American Graffiti and Chicano Murals in East Los Angeles, 1972–1978," in Bright and Bakewell, *Looking High and Low: Art and Cultural Identity.*

48. Raegan Kelly quoted in Brian Cross, *It's Not about a Salary: Rap, Race, and Resistance in Los Angeles* (London: Verso, 1994), 66. Despite such intentions by lowrider-friendly police, antagonism between lowriders and police is ongoing. Many lowriders have their own stories of being pulled over for no apparent reason. The most intense moments of conflict between police and lowriders come when popular cruising areas, such as Whittier Boulevard in East Los Angeles, are closed. See Stone, " 'Bajito y Suavecito' "; Vigil, *Barrio Gangs;* Vigil, "Car Charros"; and Robert Rodriguez, *Assault with a Deadly Weapon* (Los Angeles: Librería Latino-Americana, 1984). See also *Low Rider Magazine* 9, no. 15 (1979); and the video documentary *Low Riders: The Real Story.* In Austin the relationship is ambivalent. Lowrider clubs help the police department with charity projects like Christmas gift drives, but individual lowriders are still pulled over and still feel harassed.

49. Gradante, "Art among the Low Riders," 36.

50. Given names without surnames here are pseudonymous.

51. The larger society [tends to] label *some* minority persons,

> a priori, as "probably deviant." Thus to be young, male, and black or Chicano in white America *is* to be a suspect person. . . . To be a visible member of a population that many Anglos associate with violent crime is to evoke hostile and fearful responses. I would like to call this "ascribed deviance."

Joan Moore, "Isolation and Stigmatization in the Development of an Underclass: The Case of Chicano Gangs in East Los Angeles," *Social Problems* 33, no. 1 (1985): 2 (emphasis in original).

52. See Ronin Ro, *Gangsta: Merchandising the Rhymes of Violence* (New York: St. Martin's, 1996) for an account of gang violence at a lowrider show in Los Angeles.

53. José David Saldívar, *Border Matters: Remapping American Cultural Studies* (Berkeley: University of California Press, 1997), 119.

54. By using the term "activist," I do not intend to imply any connection between these clubs and mainstream liberalism, the American Left, or other groups that use the term. Rather, "activist" here is descriptive of those who see expressive culture as a means of social intervention rather than "just" recreation or entertainment. The lowrider clubs in question would be more likely to use an evaluative term like "positive." I have followed their usage where I feel it is appropriate.

55. Bright, "Remappings: Los Angeles Low Riders." This tendency is of course also salient in other urban expressive cultural forms. While lowriding and graffiti can both be used to mark territory, it is also a prime aim of lowriders and graffiti writers alike to "get known" throughout the city. The inherent mobility of lowriders is matched by graffiti writers' preference for decorating modes of transportation: subway trains in New York City and freeways in Los Angeles become their canvases, where everyone can see their work. See Rose, *Black Noise*. As Stewart notes in her discussion of graffiti, this mobility is in stark contrast with the sharply demarcated zones and turf battles of street gang culture, "Ceci Tuera Cela: Graffiti as Crime and Art," in *Crimes of Writing: Problems in the Containment of Representation* (Durham: Duke University Press, 1994).

56. It is also crucial to recognize that Anzaldúa's "new mestiza consciousness" cannot be simply translated to a masculine, inclusive mestizo term, as she writes quite explicitly from a woman's position and women's resistance to patriarchy and exclusion is central to her project. By looking to female lowriders as models for lowriding in the future, I am trying to imagine a lowriding practice informed by critiques like Anzaldúa's.

57. Penland, "The History of *Lowrider Magazine*."

58. One notable exception is *Lowrider Bicycle*, which does not use glamour photography, presumably since it is aimed at a younger audience.

59. Increasingly, lowrider shows also feature a "hardbody contest" or "macho man" male pageants in the interest of "equal time," to appeal to female spectators as well as the men. In lowrider publications, however, glamour photography of professional models exclusively depicts women. See Bright, "Mexican American Low Riders."

60. This story, along with various responses to the issue, was distributed on the email discussion list "Discussion/Advocacy of Human, Labor and Civil Rights," 96seradc@u.washington.edu.

61. Anzaldúa, *Borderlands/La Frontera*, 21.

62. The portrayal of the women critics as traitors invokes narratives of La Malinche, the indigenous woman who, at the time of the Spanish conquest, became Cortés's concubine and translator and thus simultaneously the mother of *la raza mestiza* and a collaborator with the *conquistadores*. A discussion of the various images of women in lowrider iconography and how they correspond to women's roles in the constituent community would be an excellent project and requires its own paper. Among the *LRM* models and car murals depicting the Virgen de Guadalupe, Tejano music star Selena, *pachucas*, and Aztec princesses, an array of ideal roles may be found: goddess, mother, daughter, lover, whore, angel, traitor. Considering the presumed gap between intellectual/academic feminist discourse and the sexual politics in which the *LRM* employees and models engage, this may be considered a case of what the anthropologist Sherry Ortner calls a "dialectic of sex," a conjuncture between two distinct sexual politics, each with its own history and debates. If this is the case, both the MEChA feminists and the models may consider themselves to be "gender radicals," in MEChA's case by challenging the objectifying practices of patriarchy, and in the models' case by launching professional, financially rewarding careers as media personalities. See Sherry Ortner, "Borderlands Politics and Erotics," in *Making Gender: The Politics and Erotics of Culture* (Boston: Beacon Press, 1996), 181–212.

63. On women's incursions into lowriding, see García, "Putting Dreams in Gear"; Mary Fears, "Chicana behind the Wheel: Women Taking Larger Part in Lowrider Onda," *El Placazo*, Fall 1996; and Jeffrey Rick, "Blue Bird," *Lowrider Magazine*, May 1997, 40–42.

Chapter Seven

Karaoke and the Construction of Identity

Casey Man Kong Lum

The word *karaoke* is a hybrid term consisting of two components: *kara* meaning empty, and *oke*, an abbreviation of *okesutora*, an adopted foreign word in the Japanese vocabulary meaning "orchestra." Taking the separate elements together, karaoke means "orchestra minus one [the lead vocal]," which refers to prerecorded musical accompaniments designed for amateur singing.[1] The word also denotes either a place (such as a bar or a nightclub with karaoke entertainment) or a machine that allows users to sing with prerecorded musical accompaniments.

To the novice, singing karaoke, especially before a group of strangers, can be a nerve-wracking experience. I can vividly remember how I felt the first time I attended a karaoke event. Not unlike many first-timers, I went through a series of complex yet memorable emotions during that first karaoke experience: the excitement, the anticipation, the anxiety, the excuses, the self-doubt, the urge to show off, and so forth. My first karaoke encounter took place in the summer of 1993 at a friend's barbecue party on Long Island. No one mentioned to me beforehand that karaoke was on the agenda. Perhaps it was not planned ahead of time. But when the grill in the backyard was slowly cooling off, our hosts, Richard and Diane, suggested that the seven or eight of us sing karaoke in the large living room. The thought that I might have to sing before a group of people, most of whom I had met

This chapter is reprinted by permission of Lawrence Erlbaum, Associates, Inc. and the author from *In Search of a Voice: Karaoke and the Construction of Chinese Identity in America* by Casey Man Kong Lum, © 1996 by Lawrence Erlbaum Associates, Inc.

only a few hours before, caused anticipation as well as anxiety. It had never occurred to me that I was even remotely close to being musically competent enough to sing in public.

In the car on our way home, I busily wondered whether everyone at the party had discovered how badly I had sung. But, almost miraculously, my wife told me how very surprised she was that I could sing so well. "So well?!" I thought. I was puzzled by my wife's compliment because I knew she actually meant what she said. Nonetheless, I began to feel better about my singing.

For several days afterward, I could not help but think about the various emotions I had experienced that evening; from the anxiety to the longing to be part of the group, and from the positive approval of an otherwise mediocre performance to the willful acceptance of such a compliment. Also, I thought about how casual and invisible karaoke as a form of entertainment and popular culture had already become to Richard and Diane, as well as to the millions of people like them in various parts of Asia, in overseas Asian immigrant communities, and, increasingly, around the world. I wanted to know more.

I begin this essay with my first karaoke experience not only because it was memorable but, more importantly, because it reflects the typical responses that people have when they first encounter karaoke. Although many people share these initial emotional responses, the significance they eventually generate through, or attach to, their karaoke experience may differ from one social context to another. In fact, it is my purpose here to examine how karaoke may be engaged in in a great variety of ways, and how varying social meanings can be constructed through the use of karaoke in different everyday contexts.

My book *In Search of a Voice*, the larger work from which this essay is adapted, is an ethnography of how karaoke is used in the expression, maintenance, and (re)construction of social identity as part of the Chinese American experience. It explores the social and theoretical implications of interaction between the media audience and karaoke as both an electronic communication technology and a cultural practice. As such, the book has its theoretical foundations resting on a nexus of three areas of analysis: namely, cultural adaptation of communication technology, audience interaction with electronic media, and the role of media in the evolution of the Chinese diaspora in the United States. In this essay I will summarize the book's findings.

Karaoke as a Cultural Practice

In conceiving karaoke singing as a cultural practice, I use the word *culture* in much the same way as Clifford Geertz's (1973) vivid conceptualization: "Believing, with Max Weber, that man is an animal suspended in webs of significance he himself has spun, I take culture to be those webs, and the analysis of it to be therefore not an experimental science in search of law but an interpretive one in search of meaning" (5). There is an intimate relationship between this interpretive anthropological conception of culture and the ritual view of communication, characterized by James W. Carey (1988) as "a symbolic process whereby reality is produced, maintained, repaired, and transformed" (23). To study communication from this perspective, as Carey pointed out, "is to examine the actual social process wherein significant symbolic forms are created, apprehended, and used" (30). Communication is an ongoing, sense-making experience whereby people, by using shared symbols, negotiate and determine among themselves the legitimate, the significant, or the sensible; that is, their social reality, the essence of their culture. Therefore, I take culture to mean the ways people interact through material, symbolic, and/or institutional means; and the meanings, values, or significance they derive from or attach to such interactions. Culture manifests, perpetuates, and transforms itself through and in communication.

Reflecting on this conception of culture and communication, we can begin to conceptualize culture at the systemic level; that is, at the level of definable, individual systems. Raymond Williams (1965) suggested that "a culture" is

> a particular way of life which expresses certain meanings and values not only in art and learning, but also in institutions and ordinary behaviour. The analysis of culture, from such a definition, is the clarification of the meanings and values implicit and explicit in a particular way of life, a particular culture. (Cited in Hebdige 1979, 6)

To study a particular culture or cultural system, therefore, is to study the symbiosis among the human, material, symbolic, and institutional elements in society in the articulation of a particular way of life. We can begin to see how karaoke singing can be analyzed as a cultural practice. Karaoke embodies a process of human interactions and practices whereby certain values, meanings, or social realities are created, maintained, and transformed as part of a culture—a particular way of life. By extension, different ways to engage karaoke represent the articulations of different ways of life. To study Chinese American immigrant karaoke, therefore, is to study how karaoke is engaged in as an everyday practice in the construction, articulation, and interpretation of those meanings that are part of the Chinese American immigrant experience. This leads to the following question: What cultural practice does karaoke embody? To address this question, I examine in the next section the social origins of karaoke in Japan.

Social Origins of Karaoke

Karaoke as a communication technology came into being under a most appropriate set of social conditions. Before mass-produced karaoke sets came along, bar patrons singing to the tunes of professional musicians had long been a tradition in Kobe, a thriving seaport metropolis in western Japan. Overlooking Osaka Bay, Kobe owes part of its glamour to its lively musical nightlife. The proliferation of cassette tape recorders also facilitated an interest in singing among many in Japan, chiefly among men in their forties and fifties (Ogawa 1990). The late 1960s and early 1970s saw a revival of *enka* (traditional Japanese songs of unrequited love) among these middle-aged men, who felt out of place with the dominating, Western-tinged popular music that was then catering to the young in Japan (Mitsui 1995). But there were efforts to reduce the cost of providing live music in drinking or entertainment establishments, where many middle-aged *enka* devotees drank and socialized after work. Experiments were attempted with several sound media, such as reel-to-reel tapes and jukeboxes.

The prototype of commercial karaoke sets did not appear until 1972. At the time the entrepreneur and musician Daisuke Inoue and his colleagues in Kobe introduced eight-track cartridges of their own musical performances without any vocal elements (Mitsui 1991). One such cartridge contained four accompaniments of two tracks each. Technically, the eight-track loop cartridge gives its users easy access to the four accompaniments: the loop goes back to the beginning as any one of the accompaniments comes to an end. Special machines were built for these tapes. Each karaoke set was built with integrated microphone inputs, an echo mechanism, and a coin slot. What was particularly unique about Inoue's tapes and the machine, as Mitsui pointed out, was that they were the first commercial accompaniment tapes and karaoke systems meant for amateur singers.

In short, the various experiments to reduce the costs of providing live music at drinking and entertainment establishments where *enka* reigned set up the immediate context for the conceptual and technological development of karaoke in Japan. Because *enka* was the most popular musical genre at the drinking or entertainment establishments where commercially available karaoke evolved, it represented the majority of the karaoke repertoire in its first decade in existence (Mitsui 1995).

Meanwhile, the Japanese love of singing helped serve as a fertile ground for the germination and spread of karaoke to other segments in Japan's population and for the development of musical genres other than *enka*. As Boye De Mente (1989) observed,

> For centuries, the Japanese have been raised on lullabies and folk songs and have learned how to sing as a natural part of growing up. Everybody is expected to join in the singing of festivals, and, at parties and other kinds of gatherings, it is customary for individuals to take turns singing. (158)

Because social amateur singing was a preexisting cultural practice in Japan, what karaoke as a technology did for that aspect of Japanese culture was make musical accompaniments more readily available to people when they wanted to sing. Therefore, karaoke as a technology is the material embodiment of a set of preexisting cultural practices, arrangements for human interaction in everyday life—or what Erving Goffman (1983) called the interaction order. Karaoke encapsulates a "web of significance," an interaction structure that, given its social and cultural roots, fosters the building and maintenance of group membership or community through participatory singing. This partially helps explain why karaoke singing has become a popular cultural form among people of all ages in Japan (Ban 1991; Nunziata 1990; Ogawa 1993b; White 1993a).[2] Similarly, Deborah Wong (1994) suggested that the love of amateur singing in certain parts of Asia helps explain why karaoke has been popular in those regions (see also "Karaoke Is," 1992).

The first U.S. karaoke bars appeared in 1983, catering to a mostly

"Asian clientele" (Zimmerman 1991, 108), which, as Mitsui suggested (1995), might have been predominantly Japanese business people working in the United States and Japanese Americans. Toward the end of the 1980s, bars catering to a more general American clientele, along with certain commercial publicists, started to use karaoke as a promotional tool (Armstrong 1992).[3] Whereas the general population in the United States is still warming up to the idea of singing karaoke in bars, certain definable cultural practices that center around karaoke have already emerged among many in Chinese America.[4] Much like their cousins in Asia, where karaoke is widely popular, many Chinese Americans from diverse social, economic, and ethnic backgrounds engage in a great variety of karaoke activities: self-entertainment, social gatherings, performance, or festive events, and so forth. Social networks are formed around a common interest in karaoke. Indeed, not only has karaoke become a popular form of entertainment for many Chinese Americans, some have adopted karaoke as a cultural practice central to their social existence.

The Karaoke Decorum

Karaoke decorum involves a set of conventions for maintaining and judging what is to be considered socially appropriate behavior in a karaoke scene. Of course, any such decorum also implies what is inappropriate or socially unacceptable. However, specific karaoke decorum varies from one karaoke setting to another. In advising his Western readers on how to behave in Japanese karaoke bars, Rex Shelley (1993) suggested,

> Karaoke bars can be great fun, or terribly embarrassing. The microphone is on a long lead and when it gets to your table, everyone must do his party thing. Do not try to get out of it by saying that you do not know any Japanese songs. They always have a few popular English language songs in the box to slap down this excuse. "Yesterday" and "My Way" seem to be the top of the karaoke western pops. Do your best. Nobody will really mind if you can't sing provided that you don't sing too long. You cannot refuse. (159)

Ogawa (1993a) offered three other tacit rules in the karaoke space of Japan: one must not sing two songs in succession, one must not sing the same song that the others have sung, and, when others are singing, one must applaud between verses and at the ending of the song.

However, although these rules of decorum may seem standard to those familiar with the Japanese karaoke scene, they are not universally observed. On one Saturday night in May 1994, I witnessed a scene at a Chinese American restaurant in a business area of Washington, DC, open for karaoke singing in the late evening. At one point in the evening, two young Taiwanese women together sang two songs in succession. With the exception of their six friends at one table, the two women were singing to an otherwise indifferent audience of about thirty people at four other tables. These other people did not applaud between verses or at the end of the songs; but spent most of the time talking among themselves. Despite this, the night went on without any noticeable unease among the participants in that karaoke scene.

The decorum observed in that Chinese American restaurant undoubtedly differs from the Japanese karaoke decorum that Shelley (1993) and Ogawa (1993a) described so matter-of-factly. In other words, karaoke scene and decorum, as with any other social space and decorum, are culture-or context-bound. Because each dramaturgical web of karaoke necessarily reflects the organizational, material, and symbolic orientation of the people who build and use it, the web may also be said to reflect what Dick Hebdige (1979) defined as the style of that group of people. Although karaoke technology varies little from one manufacturer to another, the differing social contexts embodied by the three interpretive communities of Chinese American immigrants that I studied can and do facilitate the construction and maintenance of divergent karaoke scenes and codes of decorum. By extension, they also represent and foster different karaoke experiences, practices, or styles.

From Reading to Production

I suggest that we conceptualize those people who participate in karaoke scenes not just as media consumers who read what they buy from the market (karaoke music) but as producers of the indigenized cultural products (their own karaoke episodes and performances) they at the same time consume. They are producers in two ways. First, they are the ones who plan or organize the karaoke scenes where they are also the performers and audiences. In the karaoke scenes they put together, participants can (and often do) set their own rules or plot the scenes for their own purposes, from the setting and rules of decorum to the specific music versions to be used.[5]

At another level, karaoke participants are the producers of the performances in karaoke scenes. Goffman (1959) referred to performance as "all the activity of an individual which occurs during a period marked by his continuous presence before a particular set of observers and which has some influence on the observers" (22). Performances are the actual, lived acts, events, and situations that are the substance of human interaction (Bauman and Sherzer 1989; Hymes 1975) that help establish, revise, and eventually maintain interactional structure in the context of everyday life (Leeds-Hurwitz, Sigman, and Sullivan 1995). Thus, karaoke performers—both singers and audience members—literally produce or create their scenes with each other.

There are two categories of performance in a karaoke scene. The first category is the participants' activity while interacting with others in the audience area. The second is the activity of engaging with karaoke music in the stage area. I now turn to the specific role of karaoke music in constructing the performance in the stage area.

In and of itself, karaoke music or songs are incomplete in their content—by definition, an orchestra without the lead vocal. Karaoke music is designed to be reprocessed, to be interwoven with live vocals.[6] Of course, the karaoke music producers have already set up in their studio the musical-technical parameters that their consumers can or are expected to perform within. In addition, many karaoke performers do try to sing "cover" (Fornäs 1994, 92)—that is, imitate or conform to the musical-technical specificities of the original music.

However, that is just one of a variety of ways people approach karaoke performance. To some people, singing karaoke is but one way to congregate with people whose company they enjoy. They really do not care much if they can sing well at all. As James, a young professional at a pharmaceutical company in New Jersey, told me at the end of a private karaoke party, "Since all of us are amateurs and we will never beat the pros like Lau Tak-Hwa [a popular singer in Hong Kong], we should then just do our best, be ourselves, enjoy ourselves. To express ourselves is the most important thing."

Similarly, there are people who for various reasons deliberately change the words of the songs they sing. I observed how Peter, an employee of Richard and Diane, presented himself at their party mentioned at the beginning of this chapter. It was known among the couple's employees that Richard and Diane held opposing views of Peter; Richard did not like Peter, whereas Diane did. In singing Elvis's "Hawaiian Wedding Song;" Peter changed the lines "I love *you* with all my heart" (to "I love *my* boss with all my heart") and "Promise me that you will *leave* me never" (to "Promise me that you will *fire* me never"). All the while, Peter turned to Richard, who was sitting comfortably in his couch, but who did not seem to pay much attention to what Peter was doing.

In other words, there is no telling prior to the performance how karaoke's prepackaged music will eventually sound when the music is mixed with live vocals or, for that matter, how participants will respond to the final outcome. Equally important, karaoke music is meant to be integrated as part of a larger event. It represents just one of the many audiovisual elements or semiotic resources that make up a performance on stage. People's costume, makeup, style and competence of presentation, persona, reputation, and so forth are as important as their choice of music in packaging their performance in the stage area.[7] Because the production of karaoke performances involves the audience in a live environment—unlike the production of any other traditional mass media—the cultural projects of karaoke are by nature dynamic, hybrid, and intensely indigenous. Accordingly, the interpretation of the social and cultural significance of karaoke's performances can often be difficult, unpredictable, and volatile.

Three Voices: Three Articulations of Life in the Diaspora

Three different ways to engage karaoke as cultural practices have emerged from the case studies I analyzed in my book. The first interpretive community consists of Hong Kong Cantonese immigrants active in New York's Chinatown who use karaoke as a cultural connection. But while karaoke provides them with a link to an older cultural practice, the singing of Cantonese opera songs (*yutkuk*), it also helps to transform how *yutkuk* is sung and presented in the contemporary social and technological environment. Personal musical competence is also influenced when the performer makes a shift from singing *yutkuk* with live music to singing with prepackaged karaoke music. Nevertheless, people in this community use karaoke to expand their social life worlds beyond the walls of their *yamngok se* and to create a ritual performance context for their compatriots in the Chinatown neighborhoods. In the process, they also help keep part of their musical tradition and ritual alive in the diaspora.

The second interpretive community is composed of Taiwanese immigrants active in affluent suburbs of New Jersey who engage karaoke as an expression of their wealth and social class. How they organize their private karaoke clubs and galas and their approach to their karaoke experiences reveals a conspicuous degree of corporate managerial mannerisms and a competitive drive. These might have been acquired as part of the members' professional ascents, and run parallel to the way they have been assimilated into the U.S. economic mainstream. Members of this community manage their leisure activity in a way that resembles their work in the corporate environment. The structure of their karaoke spaces also manifests a patriarchal social and moral worldview.

The third interpretive community is represented by certain Malaysian Chinese active in the Flushing area of Queens in New York City who employ karaoke as an escape mechanism. Multiply marginalized by their relatively low economic status and (for some) undocumented immigrant status, and as a minority within a minority, people in this interpretive community find solace in the communal webs of karaoke. Through karaoke, they construct a voice of their own that articulates their human condition, their alienation, their loneliness, and the absence of recognition from the larger social environment. Through imaginative uses, karaoke is transformed into a sort of therapeutic heaven where they can find relief from their isolated, humdrum routine.

Expression of Ethnicity

The diverse karaoke experiences I analyzed express the complex ethnic composition of people across the three interpretive communities. According to W. W. Isajiw (1974), a people's ethnicity "is a matter of a double boundary, a boundary from within, maintained by the socialization process, and a boundary from without established by the process of intergroup relations" (122). As minority people in multiethnic America, these first-generation Chinese immigrants have to constantly negotiate with people in the dominant society and other minorities to establish and maintain their social space and representation. They maintain a strong tie with people from similar backgrounds, especially when facing overt and covert racism or racial ignorance and hostility in the larger society. Meanwhile, regionalism plays an important role in how many Chinese immigrants define or establish their intragroup membership and bonding. Historically, many Chinese immigrants made their adjustment by staying close together for social acceptance, economic survival, political protection, and cultural familiarity.

This is not to say that people in the three interpretive communities necessarily go through the same assimilation process. Whereas many Chinese Americans remain in their social and ethnic enclaves (like many people in Mrs. Chung's community in Chinatown and Ah Maa's community in Flushing), many others acquire and maintain a pluralistic profile (like many people in John's community in New Jersey). On closer scrutiny, however, even the extent of assimilation achieved by John and his compatriots is not complete or thorough. Assimilation, according to R. T. Schaefer (1979), is the process whereby an immigrant or immigrant group acquires the traits of the dominant society and is ultimately absorbed into that society. There is no doubt that people in John's community have acquired some traits of the dominant economic structure, such as corporate managerial mannerisms, professionalism, a material lifestyle, and so forth. But they largely remain with people from their ethnic and regional background for the maintenance of their personal life world. Recall what Louis, the president of a private karaoke club, said: "Frankly, there are interactions with Americans at the professional or business level. But when it comes to personal life, it is seldom that Chinese enter the social network of Americans, or vice-versa."

One should not overgeneralize Louis's experience to represent the experience of all first-generation Chinese immigrants, but my observation of their karaoke practice and attendance to their testimonies leads me to conclude that Louis's attitude is shared by many people in the three communities. They are not ethnocentric, nor are they xenophobic—for they do interact to varying degrees with people from outside their own ethnic background at various points. As first-generation immigrants, they simply do not share the same social and life history with people who otherwise grew up in the dominant society and the different ways of life it embodies. Therefore, it is understandable that certain first-generation immigrants cannot, do not, or will not fully identify with people in the dominant society at the social (personal life) level. Accordingly, there is reason to wonder if assimilation—in Schaefer's (1979) terms—can ever be accomplished.

People's regional backgrounds also play a role in their musical choice, which creates boundaries of its own. For example, people in Mrs. Chung's community use mostly Cantonese opera and popular songs from Hong Kong. Coming from a regional and linguistic background in Taiwan, people in John's community use mostly Mandarin and Taiwanese songs. The presence of Japanese-language or Japanese-tinged music (such as *enka*-style Taiwanese songs) also reveals a degree of Japanese cultural influence in the lives of some of these people. It is a trait largely absent from the other two communities. Finally, the karaoke music used by people in Ah Maa's community tends to be a mix of Cantonese, Mandarin, and Taiwanese songs. Cantonese, Mandarin, and Taiwanese are part of these people's language community in Malaysia; media products from Hong Kong and Taiwan encode these three Chinese dialects.

At the level of East-West cross-cultural consumption of karaoke music, John's community seems the most pluralistic of the three. On the average, about one-sixth to one-fifth of the songs performed at the large-scale karaoke galas are English-language (notably American) songs. To the contrary, I rarely hear members of the other two communities sing English-language songs. Of course, this fact alone does not suggest that people in these two communities cannot sing such songs. It may be just a matter of their own musical preference, partly defined by their personal and regional experience although the nonselection of English songs is widespread in these two communities.

In short, the karaoke practices of the three interpretive communities, including their specific choices of music, are expressions of the distinct ethnic and regional backgrounds of the people as they construct and maintain social membership. The fact that people in these three communities socialize mostly with others of a similar regional background is indicative of the important role that regionalism plays in the construction and maintenance of Chinese identity, that Chinese are not just Chinese, but Chinese from "where." Of course, it is important to note that the experiences of these first-generation immigrants cannot and should not be generalized to represent how their offspring in the diaspora might construct, maintain, or negotiate their ethnic identity—for the latter are acculturated in a different society, a different region.[8] Moreover, they might also have views of their parents' homeland or cultures that are different from what their parents have in mind or practice.

Expressions of Class

The karaoke practices of the three interpretive communities are also expressions of the people's divergent economic experiences, at both the inter-as well as intracommunity levels. At the intercommunity level, the three interpretive communities represent three economic experiences or classes in Chinese America. People in John's community belong to an upper middle to upper class; they are economically accomplished and fairly well assimilated into the U.S. mainstream. Members in Mrs. Chung's community, in contrast, belong to a middle to lower middle class, living comfortably but remaining mostly in the economic enclave of New York's Chinatown. Finally, people in Ah Maa's community, with the exception of a few entrepreneurs, come from a working-class background and survive in a mostly underground economy, some by living at the minimum-wage level.

The economic distinctions and lifestyles of the three groups are manifest in how they materialize their karaoke scenes. These people's karaoke webs of significance are spun and suspended in three different social spaces. One can easily discern, for example, the extravagance embodied in the karaoke scenes in John's community as compared to the material simplicity of the karaoke scenes where people in Ah Maa's community mostly congregate. These webs do not intersect, for people do not readily cross between each other's karaoke spaces as a matter of course. People in these three interpretive communities do not have any genuine, sustained interclass social interaction.

In certain cases, some people in the New Jersey community were quite unwilling to be in any karaoke scenes that did not match their own. These people have better karaoke facilities at home than are found in typical Chinese restaurants. One such member saw it this way: "Those whose equipment is not as good as the clubs [that open to the public] may want to go to those clubs." According to this person's view, the karaoke facility that one has (or does not have) becomes a class symbol.

Asked if he had gone to karaoke clubs in Flushing, Queens, another affluent club member responded, "Flushing? No, it's too far away." But when I pressed further by suggesting that there were a few karaoke clubs in New Jersey that were open to the public, this man said,

> I heard *the level of the people going to those clubs* [a pause here] . . . I don't know. . . . They smoke in the club, *wuyen changch'i* ["full of bad or filthy smoke and fume"]. Why are we going [there]? If we want to chat with our friends, it's comfortable to do so at home. . . . We haven't been to any of those clubs. (Italics added)

Notice that this informant paused after his reference to "the level of the people going to those clubs." He quickly shifted to smoking and cigarette fumes as his reason for not going to "those clubs." But the aphorism *wuyen changch'i*, referring to bad smoke and smell at the sensory level, also is a value statement suggesting how a place is filled with social ill and filth. The aphorism might have been used connotatively at both the sensory and the social-valuative levels.

People in Ah Maa's and Mrs. Chung's communities are not entirely immune to this kind of social self-selectivity (or exclusivity), although I did observe on several occasions that people in the lower economic class seemed to be receptive to mixing with people from a higher economic class. At Ah Maa's first karaoke birthday, one of her friends brought her male employer and conspicuously introduced him with the words "He is a boss." People who overheard the introduction invariably turned their heads to the man. This introduction and the reaction it caused seemed to indicate a degree of class consciousness among people in this community, including the woman herself. But people at the party did not appear to reject the boss, although they initially maintained some distance from him. It was halfway into the night before he was able to blend in with a small group of guests active in singing karaoke. I cannot help but keep wondering how, without the woman as a guide, he might have otherwise fared with this crowd, many of whom survived on minimum wages from their own employers.

It is also important to note that, at the intracommunity level, members may have varying economic experiences. How individuals within a community engage karaoke certainly expresses their economic attainments and social status as members of that community. However, my comparative analysis of the three cases indicates that intragroup competition exists and that it is most pronounced in the affluent interpretive community in New Jersey. Karaoke along with other signifying objects such as houses and music teachers establishes one's status within the community. This kind of intragroup comparison or competition does not seem to manifest itself in the other two interpretive communities.

Expressions of Gendered Practice

A gendered interpretation of technology enables us to conceptualize technology use as a social practice that reflects gender relations. Certain karaoke practices I observed in my ethnography indicate such gendered practices. For example, a woman club member once observed,

> Guys tend to like musical instruments more. If the husband doesn't want to buy [a karaoke machine] for his wife to sing. Maybe one or two. Guys like the wires, or things like that. We [women] tend to reject them. Our friends are mostly like that. When we have a gathering, the women usually get the tea and chat or something like that. When the guys are ready with the setup, we then go to sing. It's always the guy to initiate: "Let's sing karaoke."

Interestingly, this woman's experience echoes the experiences of Tim O'Sullivan's (1991) female informants, whose husbands tended to be the ones to decide on the purchase of family television sets (cited in Moores 1993). Moores (1993) and Morley (1986) also found a tendency of male dominance in, respectively, the uses of early wireless and television remote control devices in domestic environments. In references to O'Sullivan's finding, Moores (1993) observed, "No doubt, this can be explained by [men's] control over large items of household expenditure, but it is related, in addition, to the connections between masculinity and gadgetry" (89).

Of course, one should be careful not to generalize the one woman's observation to the experience of all other female members of the community.[9] But the observation did come from within the community and is purported to represent the experience of a large network maintained by the woman's family. Similarly, my field observations indicate that the men in the New Jersey community, as well as those in the other two communities, are more in control of the machine. For example, Mr. Hau and particularly Bill (Mrs. Chung's son) were always the ones to set up the karaoke *yutkuk* street parties. John had undisputed control of his karaoke ballroom. Calvin conceived and managed the entire technical setup for a couple of galas after he became the president of his club. Carrying over the technical role in his club to other karaoke scenes, Ah Ting often played disc jockey at Ah Maa's birthday parties.

Other nontechnical activities also reveal the gendered nature of the three communities. The marriage requirement of private karaoke clubs in John's community essentially assumes, and is put in place to enforce, traditional family structure and, by implication, the heterosexual sociomoral order. The story of a karaoke-related affair that had been circulating among many people in this community also embodies such a worldview, which is also a male-dominated one. The married man in the narrative that was told to me repeatedly was the central character, while the other woman (also variously identified as simply "some woman" or "a woman") was portrayed on the sideline. The story never mentioned what might have happened to the man's wife during or after the alleged affair. The women in the story thus appear as passive, subordinate characters in a male-centered social and moral theater. Similarly, one may recall Calvin's Elvis act onstage, with the dancing girls playing the supporting roles.

In contrast, the action of Susan (in Ah Maa's community) in the late-night melodramatic excursion in the basement of the Southeast Asian restaurant expresses another kind of gendered power relations. The larger social world in which Susan lives is one of patriarchy. But although Susan was the romantic (or sexual) target of Denny and Ah Dong's flirtations, and thus their perceived or assumed subordinate figure, she was not truly a passive player in that karaoke episode. If she ever was the hunted, Susan certainly outmaneuvered her male pursuers. One may even suggest that Susan helped facilitate the episode so that she could enjoy the pleasure of the men's pursuit before discarding them. Susan's action was a symbolic gesture of her defiance of the passive, subordinate female gender role.

I am not suggesting that male dominance is absent from Susan's interpretive community, nor am I hinting that her compatriots enjoy symmetrical gender relations. After all, Ah Dong (and Denny) initially appeared to have assumed their lead and control. But Susan's action in that karaoke episode does reflect a degree of looseness in the social, interactional structure of her community's karaoke spaces. Unlike the karaoke webs in John's community (the galas, the music lessons, the monthly club meetings), which are themselves extensions of entrenched social institutions and norms (such as family, marriage, and corporate culture), those in Susan's community are, by the nature of the community itself, transitory. They are temporary, brief interactional structures where hegemonic social norms seem to be harder to enforce. The dramatic elements encoded in karaoke music and videos make the rules or patterns of human interaction in such a transitory space even more unstable and, to a certain extent, unpredictable. From Malaysia, Susan is out of reach of the social birthplace in which she is well connected and, from another perspective, constrained. As a global migrant, she can defy, to a certain extent, traditionally entrenched gendered roles while traveling in and through transitory social spaces both geosocially (Flushing, Queens) and dramaturgically (Ah Maa's karaoke birthday parties).

In short, the divergent ethnic, social, economic, and gender frames of reference for people in the three interpretive communities—and the diverse karaoke experiences they engender—reflect the multiplicity and complexity of Chinese American culture. Although it is much easier for some people to view Chinese American as a homogeneous entity, such a view conceals much that is worth considering. At some level, it may also be easier and more convenient and comforting to view Chinese Americans (or Asian Americans as a whole) as a singular, model minority. However, such a view offers a picture of success, achievement, and happiness in the lives of some, and it shows little of how many others must confront hardship, neglect, and disappointment as they struggle in their everyday lives in the diaspora.

NOTES

1. The word *karaoke* has been translated literally as *empty orchestra* (e.g., see Feiler 1991, 51; Shelley 1993 159). But according to Toru Mitsui, in the original Japanese conception of karaoke, *kara* carries the connotation of *without [the voice]* instead of the literal meaning of *empty* (1995). Mistsui has written on the history of karaoke in Japan and, at this writing is the president of the Inter-

national Association for the Study of Popular Music. Yoshio Tanaka (1990) similarly referred to karaoke as sound tracks minus the lead vocal.

2. Merry White (1993b), a Boston University sociologist and a research fellow at Harvard University's Edwin O. Reischauer Institute of Japanese Studies, added yet another interesting explanation for why karaoke has become a popular form of entertainment in Japan. She suggested the following:

> Japanese don't invite people home. A karaoke bar or club is a home from home. It's always important that there is a place where one can entertain friends. Home is not a place to entertain guests, especially if they are not one's own family. It's been said that Japanese don't invite friends over because houses in Japan are small. But that's not really it. It is about a sense of a home, a place that is inviolable. It is a private space where one can indulge in total freedom without having to put on special masks.

Karaoke offers the Japanese an important alternative space for maintaining social relations with people from outside the family.

3. Around the same time, the consumer electronics manufacturer Pioneer marketed the world's first laserdisc karaoke systems for the consumer market (Ogawa 1993a).

4. Karaoke is also popular in many other Asian American communities. For example, Wong (1994) studied the karaoke experience of certain Vietnamese Americans in the Los Angeles area, and Robert Drew's (1994) dissertation touched on certain aspects of Japanese American karaoke bars in and around the Philadelphia area. But because Asian America is a complex sociopolitical space, and because karaoke as a serious scholarly subject is a relatively uncharted area, I delimit my discussion to only certain social and cultural aspects of the Chinese American Karaoke experience.

5. For legal or financial reasons, karaoke music producers normally do not use the master sound tracks by the original artists. Instead, they hire their own performers to reproduce or imitate the original music. As a result, there can be as many versions of one song as there are software producers. Some people who are used to one particular version of a song may not like or be able to sing renditions by other producing groups.

6. One can watch a music video on a karaoke laserdisc with its vocal track on, but that is an act of watching video, in much the same way as one watches any prerecorded video or music video on MTV. It does not constitute a karaoke scene.

7. In his study of a sample of karaoke scenes in Taiwan, Ringo Ma (1994) suggested that a performer's reputation and regional background could play a role in determining how the audience would react to the performance.

8. Based on a literary analysis, Victoria Chen (1994) provided an insightful

discussion on how certain second-generation Chinese Americans resolve some of the identity dilemmas they face.

9. It is important to note that the woman's statement does not refer to whether the women in her community are competent in handling technical gadgetry. I observed some of them capably operating relatively sophisticated electronic appliances at home. To a certain extent, I suspect that Ann Gray's (1986, 1992) observation of her subjects' "calculated ignorance" (cited in Moores 1993, p 94) can help us understand this aspect of the discussion. Gray discovered that some of the women she talked to did not want to learn how to operate their family's video recorder because that alleviated them from having an additional chore in the household. More research will have to be done to determine whether certain women in the New Jersey interpretive community maintain this calculated ignorance.

WORKS CITED

Armstrong, L. "What's That Noise in Aisle 5?" *Business Week*, June 8, 1992, 38.

Ban, S. "Everyone's a Star." *Look Japan*, April 1991, 40–42.

Bauman, R., and J. Sherzer, eds. *Explorations in the Ethnography of Speaking*, 2d ed. Cambridge: Cambridge University Press, 1989.

Carey, James W. *Communication as Culture*. Boston: Unwin Hyman, 1988.

Chen, Victoria. "(De)hyphenated Identity: The Double Voice in *The Woman Warrior*." In *Our Voices: Essays in Culture, Ethnicity, and Communication*, ed. A. Gonzalez, M. Houston, and V. Chen. Los Angeles: Roxbury, 1994.

De Mente, Boye. *Everything Japanese*. Lincolnwood, IL: NTC Publishing Group, 1989.

Drew, Robert. "Where the People Are the Real Stars: An Ethnography of Karaoke Bars in Philadelphia." Ph.D. diss., University of Pennsylvania, 1994.

Feiler, B. S. *Learning to Bow: Inside the Heart of Japan*. New York: Ticknor and Fields, 1991.

Fornäs, J. "Karaoke: Subjectivity, Play and Interactive Media." *Nordicom Review of Nordic Research on Media and Communication* 1 (1994): 87–103.

Geertz, Clifford. *The Interpretation of Cultures*. New York: Basic Books, 1973.

Goffman, Erving. *The Presentation of Self in Everyday Life*. New York: Doubleday Anchor, 1959.

———. "The Interaction Order." *American Sociological Review* 48 (1983): 1–17.

Gray, Ann. "Video Recorders in the Home: Women's Work and Boy's Toys." Paper presented to the second International Television Studies Conference, London, 1986.

———. *Video Playtime: The Gendering of a Leisure Technology*. London: Routledge, 1992.

Hebdige, Dick. *Subculture: The Meaning of Style*. London: Methuen, 1979.

Hymes, D. "Breakthrough into Performance." In *Folklore: Performance and Communication*, ed. D. Ben-amos and K. Goldstein. The Hague: Mouton, 1975.

Isajiw, W. W. "Definitions of Ethnicity." *Ethnicity* 1 (1974): 111–24.

"Karaoke Is Popular in Chinese Communities around the World" (in Chinese). *China Times Weekly*, American ed., December 27, 1992, 8–13.

Leeds-Hurwitz, W., S. J. Sigman, and S. J. Sullivan. "Social Communication Theory: Communication Structures and Performed Invocations: A Revision of Scheflen's Notion of Programs." In *The Consequentiality of Communication*, ed. S. J. Sigman. Hillsdale, NJ: Lawrence Erlbaum Associates, 1995.

Ma, Ringo. "Ethos Derived from Karaoke performance in Taiwan." Paper presented at the eighty-fifth Eastern Communication Association Conference, Washington, DC, April 1994.

Mitsui, Toru. "Karaoke: How the Combination of Technology and Music Evolved" Paper presented at the sixth International Conference on Popular Music Studies, Berlin, Germany, July 1991.

———. Personal communication. April 10, 1995.

Moores, S. *Interpreting Audiences: The Ethnography of Media Consumption*. London: Sage, 1993.

Morley, D. *Family Television: Cultural Power and Domestic Leisure*. London: Comedia, 1986.

Nunziata, S. "U.S. Karaoke Outfit Plans to Enter American Market. *Billboard*, January 6, 1990, 12, 88.

Ogawa H. "The Karaoke Way." *Pacific Friend*, October 1990, 32.

———. "Karaoke in Japan: A Sociological Overview." Paper presented at the eighth International Conference on Popular Music Studies, Stockton, California, July 1993a.

———. "Unstoppable Karaoke." *Pacific Friend*, May 1993b, 17–21.

O'Sullivan, Tim. "Television Memories and Cultures of Viewing, 1950–1965." In *Popular Television in Britain: Studies in Cultural History*, ed. J. Corner. London: British Film Institute, 1991.

Rakow, L. F. "Gendered Technology, Gendered Practice." *Critical Studies in Mass Communication* 5, no. 1 (1988): 57–70.

Schaefer, R. T. *Racial and Ethnic Groups*. Boston: Little Brown, 1979.

Shelley, Rex. *Culture Shock! Japan*. Portland, OR: Times Editions Graphic Arts Center Publishing, 1993.

Tanaka, Y., ed. *Japan As It Is: A Bilingual Guide*. Rev. ed. Tokyo: Gakken, 1990.

White, M. *The Material Child: Coming of Age in Japan and America*. New York: Free Press, 1993a.

———. Personal communication. October 12, 1993b.

Williams, Raymond. *The Long Revolution*. New York: Penguin, 1965.
Wong, Deborah. "I Want the Microphone: Mass Mediation and Agency in Asian American Popular Music." *Drama Review* 38, no. 2 (1994): 152–67.
Zimmerman, K. "Can Karaoke Carry a U.S. Tune?" *Variety*, August 5, 1991, 1, 108.

Chapter Eight

Sound Effects

Tricia Rose Interviews Beth Coleman

In this dialogue, Tricia Rose, author of *Black Noise: Rap Music and Black Culture in Contemporary America*, one of the earliest scholarly analyses of hip hop music and culture, interviews the electronic musician Beth Coleman, who DJs under the name M. Singe, about the gendered nature of music production and DJ-ing; technology and aesthetics; and the "diva" history in black popular music.

Tricia Rose: Your musical activity is interesting in a number of ways. In our conversation, I'd like to explore the interplay of gender, race, aesthetics, and technology. But before we get into that, let's talk about how you started DJ-ing and how you came to found the SoundLab collective.

Beth Coleman: I was working as a writer, I hadn't started graduate school yet, and I was looking for a community in a weird way. (It seems like a funny thing to say—can you really be looking for a community?) I had been hanging out with a crew of people in the early nineties who were doing "experimental parties"—events where you'd have four or five different DJs all playing at once. This is when I started working with Paul D. Miller, a.k.a. DJ Spooky, and others of the "illbient" New York scene.[1] A lot of these parties happened on the Lower East Side, everyone was welcomed, and the party line was, "It's open, it's flow." But it was always the men playing during the party and the women cleaning up at the end, as if these things were set in stone. I wasn't really with that. I didn't want to participate within those boundaries, but I also didn't feel like being by myself writing all the time.

I started playing my first sets when my friend Howard Goldkrand and I started SoundLab in 1994. It was our party—we ran it, we set it up, we designed it. Every SoundLab event ended with a "jam" and this was a great opportunity to practice playing live without having to run a whole set. When I started playing, I got to play a little at the beginning or the end of an evening. But this isn't always the way to get to play full sets, because they'll certainly let anyone, particularly female DJs, play the opening or the closing until the cows come home. To play a whole set, you have to make some kind of jump—grab some space as it were.

After these parties became popular, the Kitchen [a New York City performance space] called me up and said, "We're doing an all-women's music night. Do you want to play?" And I asked myself, "Now, why is it that I only get called when it's an all women's night?" But I said yes, because I had to play, and that's really what it came down to. So that's how I started. And now, several years later, I am invited to places like Vienna, Istanbul, and Berlin.

Tricia Rose: How did you plan parties and set up events? Where did you get the equipment?

Beth Coleman: It was an elaborate process. We would start out at around three in the afternoon, go to pick up all the equipment, and then move it all. We borrowed a lot of it from friends; some of it we rented. We'd cobble things together by renting pieces of a sound system and borrowing somebody's decks.

Once we had the equipment, the next thing we did was cover the floor with vinyl and tape it down because dancers were also using the space for rehearsal and we had to make sure that the floors didn't get damaged. Then we set up the equipment and put in a visual installation. There were two or three people, tops, doing all this setup. I had to learn about all this equipment because there was so much to do and hardly anybody else to do it. I carried it; I plugged it in; I made sure the signals showed up on the mixing board; I learned about eq-ing; I learned about sound systems; I learned about turntable technology.[2]

The event would go from eleven o'clock until about five in the morning. Then we'd break all the equipment down, roll the floor up, and leave, at around nine or ten. It's about an eighteen to nineteen-hour event, nonstop.

Tricia Rose: Do you practice putting together sets at home, or lay out whole sets in advance?

Beth Coleman: If I'm really nervous about a gig, I'll basically lay a set out because I don't want to have to kick myself afterwards for not being prepared enough. But those aren't necessarily my best sets; my best sets are usually when I've got a group of records that I'm familiar with, I've been playing them in and out, and I feel loose enough that I can go with instinct and put something together. So that's the best for me and that's also experientially the most fun.

Tricia Rose: Why did you choose the name M. Singe? What is your interest in the name and how does it represent what you're about?

Beth Coleman: It came from a nickname in college. We were reading *The Signifying Monkey* and a friend of mine starting calling me "the little monkey."[3] "Singe" is a play on this nickname; it is French for monkey. But in English, "singe" also means to burn, or incite, which was the kind of flavor that I played. I try to incite things, to blow it up in some ways. In fact, I play a lot of things with explosive sounds in them.

When I started in 1995, I spun a kind of roughneck style of dub and jungle—lots of hard-kicking break beats, deep bass, and killing noise—Prince Far-I, Christoph de Babylon, Merzbau, and that ilk. Recently, I've been playing a more stealth electro sound (Underground Resistance, Basic Channel, DJ Assault)—lots of fat, dancy beats and dirty words. I don't play one style of music because I don't think there is so much of a difference between drum and bass, hip hop and hard-core that you can't play them all together. There are definitely people and many DJs who argue that these things are genre-specific and you're not supposed to try to blend and cross-pollinate. But my style is more than a particular rhythmic or dance style.

Being a DJ is like being a cipher of the most radical kind. In each place you play, the audience takes a new face and so do you. Sometimes I make subtle shifts between rhythm patterns—going from a jump-up drum and bass roll to some funky electro booty-shaking business without losing the sonic thread. Sometimes I make a more radical shift between styles and venue—I'll play a moody, ambient set for a performance with Butch Morris and the next night play on a big system for a thousand kids. Not everyone wants to play in such varied contexts. I imagine these variations to be like the "morphing" power possessed by Anyanwu, a character in the Octavia Butler novel *Wild Seed,* who changes her identity to meet each situation. But no matter what form she takes, there is still some kernel of "herself" she holds on to that preserves her sensibility. DJ-ing is shape-shifting like that.

Tricia Rose: SoundLab and Cultural Alchemy, the artist collective, have a reputation for being sort of intellectual musicians, and you are working toward a Ph.D. in comparative literature. What is the relationship for you between this kind of complex analytical language and the music? And what would you say to someone who might ask you why you need such a complex verbal apparatus to describe sound? Shouldn't sound be able to stand on its own terms?

Beth Coleman: Well, a lot of the analytical materials that I've been reading and thinking about, in terms of my academic work, have to do with where things break and then turn back on or reflect themselves. I do think that's one thing that you hear in the music that I play. It's a beauty of speed: something speeding up and then stopping or, literally, a thick cut going to an entirely different soundscape—different but somehow "right." You accumulate speed by halting. It's a phenomenon of getting to the point where you're pushing the system and pushing the system until it breaks. But it isn't enough to simply "cause wreck"; there has to be some reflection about what's going on. So this is what I've pursued as a writer and sonically.

Tricia Rose: So you have a theoretical interest in the relationship between kinds of continuity, rupture, and self-reflexivity that you can explore through music. What, then, do you think about technology? Does the technology enable you to reach other creative goals or is mastering/manipulating the technology a goal in itself?

Beth Coleman: The work that I've done has all been with certain technologies—turntables, mixing boards, digital samplers, or different digital audio programs like ProTools or Logic. With this technology you have to know what you're doing because if you don't know how to do certain commands, or how to initiate certain programs, the process can seem more intimidating than picking up a guitar to write a song. But my experience with technology is that I find it incredibly freeing. Once you understand some of the structures and limitations that you are working with, then of course you try to figure out how to get around them. Sometimes, sitting in the studio you can come upon these accidents, but you think, "oh that's the bomb" and then you just flow like that. (Flow, in this case, includes working like hell to get the mix tight after the initial point of inspiration.)

One of the properties of a sampler is it gives you the ability to make an absolute reproduction; you can choose to have no distortion. You can take a Led Zeppelin riff and sample it directly and you get absolute clarity in reproduction. Or sometimes the music itself disappears because you get to work on such a tiny level, tweaking it, changing its sound around, adding a phrase to it; the sampler becomes the instrument that is being played. Some of this music you'd be bedeviled to try to play live, because you can make these weird breaks and sutures with these types of machines that are virtually unplayable.

I think the music is particular to the technology. The technology allows us to create new musical structures, like the break beat where you go back and forth between the drum solo on two records. I find that this structure—this kind of flow and cut—which is a very black, a deeply black structure, to be really resonant in this technology.

Tricia Rose: It seems as though you believe that technology is a means to an end and sampling enables a kind of aesthetic expression that has its origins elsewhere. So the cuts, breaks, and collages we hear in music such as hip hop are enabled by recent technological developments but are not the result of these developments. In this way, the technology becomes a vehicle for already existing aesthetic desires. And yet the technology also stages these musical ideas and can therefore be understood as having an impact on them.

Beth Coleman: Yes, for example, much of funk music was about people's fantasies of turning into robots and machines, before these fantasies became more easily manifest through technology. Now you can be a robot by putting your voice through a vocoder like Dr. Dre and Tupac's in their Mad Max–inspired music video, "California Dreamin'."

Tricia Rose: It's funny that you mentioned Dr. Dre and Tupac in reference to the vocoder because I thought of Roger Troutman, of "I Wanna Be Your Man" and "Computer Love" fame. Perhaps this has to do with the permanency of late adolescent musical memories!

Beth Coleman: Oh yeah, that's what Dr. Dre and Tupac were riffing on, that's what they were sampling. It's become a classic sound.

Tricia Rose: Troutman's use of the vocoder takes place in a different moment, and enables an entirely different narrative, sonic aesthetic, and type of human connection than Dr. Dre. Troutman is coming out of the mid-1970s when a more traditional set of assumptions about community, romance, intimacy, and connection was operating, and he uses this technology to articulate these more traditional narratives. For example, "I Wanna Be Your Man" is basically a traditional R&B tune about romantic desire narrated by this high-pitched electronic voice. It has the effect of increasing the sense of longing in his voice. Troutman is using the vocoder on top of what some might consider to be pre–high-tech narratives of "whole" "unmediated" human relationships. On the other hand, I would say that the opposite is true for Dr. Dre. He is using the vocoder (and sampling equipment more generally) to narrate mediated and fractured relationships and to enhance the effects of a more apocalyptic vision.

Beth Coleman: That song and that video in particular are like sample culture run amok. They're sampling a movie, *Mad Max: Beyond the Thunderdome,* with Mel Gibson and Tina Turner, and they're sampling a sound, a sound that was crucial to early hip hop but that sounds retro at this point. With these music technologies it becomes difficult to determine where one thing ends and another starts.

Tricia Rose: In other words, technologies foster a kind of creative expansion that breaks or gets around boundaries. But it is still true that we need some distinctions between reality and fiction for political reasons. We do not want to say that there is no distinction between actually shooting folk and using the notion of murder as a metaphor to describe one's musical prowess. I think this is an important distinction because people leap very quickly from supporting aesthetic freedom to imagining that artistic production is without political consequences. This is a deeply complex problem for hip hop as it approaches its twentieth year of commercial success. It can no longer be thought of as a form that simply reflects social realities. It also produces them. I think it would be very difficult to argue that some of the more disturbing elements in hip hop are not assisting in the production of dysfunctional and anti-social identities for black people. If it did not have such potential, then how could it possess the possibility to produce social change or enable resistance? It has to be enabling—in both progressive and regressive ways—otherwise the agency that drives the notion of cultural politics is gone.

Beth Coleman: I think that this is a very strong argument you're making and it's not necessarily a popular one.

Tricia Rose: Yes, it is not popular among hip hop supporters because right-wing cultural voices have deployed this argument against hip hop, creative black expressions, and black people in general. This argument is not used as a way to seriously consider the complexity of cultural forces in society, but to dehumanize black people and sever the existence of black suffering from its roots in a national system of economic and racial inequality.

Cross-Fade

Tricia Rose: What types of music and sounds do you sample?

Beth Coleman: In the middle of the "Stereophonic Retina Mix," an ambient piece SoundLab created for the Whitney Museum [of American Art in New York City], there was a sample of Pigmeat Markham. Markham was an early vaudeville performer; he did classic black comedy and a number of his recordings survived. I think you can hear the whole history of vaudeville in ten or fifteen seconds of listening to his voice. So I enjoyed throwing that in the mix and seeing people respond to it.

Tricia Rose: Let me ask you about the recognition of cultural history in a sampled context. On a public radio program, I exchanged comments with a music professor who expressed enormous enthusiasm about the intertextual possibilities of new musical technologies. While there's a part of me that completely supports his reasoning, I find myself wondering about how aesthetic intertextual reference works when most people have little access to black cultural history? In other words, what about people who don't really know the history of Pigmeat Markham? Is this history in the sound of his voice or his story, or do you have to have an enormous degree of cultural literacy to be able to understand it?

Beth Coleman: I was on a panel with David Henderson and Steve Canon [of the 1960s poetry collective UMBRA and more recently, the Gathering of Tribes] and a young poet. This poet felt strongly about upholding "traditional black history." He said hip hop was hell's music and that you shouldn't be able to mix and cut because people won't know what it is you're talking about. The best I could say was we have to fight for people to have as much space as we can and we also have to teach each other so we don't forget.

Tricia Rose: The fight for freedom of cultural expression is an enormously important one. But we can't make a political statement and argue that it has political resonance if there is no collective recognition of the statement's politics. It is at this point that cultural literacy becomes of the utmost importance.

Beth Coleman: Yes, but it's a dangerous thing to assume too much about your audience. I don't believe I have to say, "Here's the mix for the museum and here's the mix for my black brothers and sisters."

Tricia Rose: Right, this is a difficult issue—who do you imagine your audience to be? And I am not sure the general Whitney Museum audience has any idea who Pigmeat Markham is anyway! Since he is a vital part of a black folk tradition there are many black people who have heard him or at least of him. But younger black people may not necessarily have heard of him given the difficulty of maintaining collective black cultural literacy in a nation that does not at a fundamental level value black popular traditions. This is not to say that you only sample something people already know—that is a vicious circle that only shrinks the referential domain.

Beth Coleman: So then everyone has to sample Sting?

Tricia Rose: You're talking about that Puff Daddy song ["I'll Be Missing You"], the one where he uses a popular Sting sample ["Every Breath You Take"], right? This seems quite different from sampling Markham to me. Pigmeat Markham was an underground critic of American society. In contrast, I would not say that Puffy is claiming he's making a political, cultural recuperation of the history of black people in Sting's sound, so it's a very different move.

Beth Coleman: It's not different.

Tricia Rose: Oh, I think it's a completely different move. There are critical distinctions between Puffy's frequent celebration of money and luxury commodities and someone who claims they're making a politically resistant statement in such a market-driven society.

Subsonic

Tricia Rose: Let's talk about gender a bit. Why do you think that women are still so enormously marginal in the electronic and technological aspects of music production?

Beth Coleman: There have always been women in hip hop, but always in smaller numbers. And of course, there are more women DJs now than there were women in the indie-rock scene. I believe that more women are going to be "naturally" drawn to electronic music than we've seen before. This is partly a sign of the times because electronic music *is* the youth culture medium of our time. Even now, at the beginning stages of electronic music, there are already more women participating in this form than there were in early rock n' roll and early hip hop.

Tricia Rose: But I'm not comparing degrees of marginalization. I am making a broader point, which is that overall, numerically speaking, music production is male-dominated, particularly when it's more technologically mediated.

Beth Coleman: In the studio sometimes you learn things directly; sometimes you pick things up because you're around. This combination of being around and being inconspicuous enough, like being someone's younger brother, fitting in the woodwork, and picking up information, is crucial. Many females don't get a chance to be someplace, be quiet for a while, listen, watch—they don't get to have this kind of osmosis learning. They have to learn by direct instruction or hire engineers, who come in and dominate the production process. It's the same with DJ-ing. Male DJs often sit around the studio with their boys, and when the DJ takes a break, one of his friends can have his first opportunity to see what it means to crossfade. This sort of casual learning relationship is often not available for females because the studio is a very stratified space. And also people tend to hang out at the studio until late at night, which may also be a problem for young women.

Tricia Rose: What you're really pointing out is the complex ways that a certain kind of patriarchy operates. It is not really a matter of women being excluded from the technology necessarily; what they lack is a casual way into this very localized culture of informal apprenticeship where they can be "inconspicuous," as you put it. This apprenticeship takes place especially in a homosocial world.

Imagine a sixteen-year-old girl who just happens to be graced with more developed breasts, trying to fade into the woodwork among eleven sixteen-year-old boys sitting around experimenting with the technology. Adolescent forms of male sexual objectification of young women can be very disabling to female freedom of expression.

Beth Coleman: But there are also ways women acquiesce to this behavior because to act as if this just goes on and women have no agency is crazy. And also to believe that these men are all only sexist is also not true. So there also has to be individual initiative.

Tricia Rose: Yes, we do want to avoid overgeneralizations and recognize strategies to get around these obstacles. But there are many factors—beyond individual behaviors—that contribute to the limited access that young women have to these spaces. For example, family and societal expectations of young women frequently have a stultifying effect on their creative development, especially in nontraditional outlets such as DJ-ing. Parents are far more anxious about how their daughters represent themselves and who they spend their time with than their sons. All of this helps to close down creative spaces in general, but definitely limits access to the sorts of spaces you mentioned earlier. This is a problem few seem to be able to address—the difficulty for women to create large pockets of collective creative production. Literature is one of the few places where women can be creative. Why? Perhaps because it is a solitary enterprise?

Beth Coleman: Or you just have to be the diva.

Tricia Rose: Perhaps the diva is a response—a "feminine" performance of power and mastery in spheres that hinder female collectivity.

Beth Coleman: And while I have respect for singing and the voice because it's a crucial part of black music, women shouldn't be limited to singing.

Tricia Rose: Women are channeled toward singing as one of the only ways for them to be creative in the music industry. (Strikingly, like literature, it is a primarily soloist domain.) The kind of masculine power associated with mastering technology is not allowed them.

I would like to hear your take on what may be an impossible question. When you have an environment in which women are relatively isolated and marginal, how might this impact cultural production in general? It would be virtually impossible to reproduce aesthetic traditions without some form of collective identity and spaces in which to create. One African American in Des Moines is not likely to do a lot of black cultural reproduction all by him or herself. I think it's very interesting that progressive cultural thinkers are generally comfortable talking about racialized aesthetics, but shy away from the possibility of gendered aesthetics, even those produced by social context.

Beth Coleman: We can say black music or black art without flinching, but I still think "women's culture" sounds like you're talking about white women.

Tricia Rose: Yes, it does. And that puts black women in an especially difficult position.

Beth Coleman: I'm not sure how to address that question because the discussion invariably leads into one about exemplary individuals and not a collective of women. DJ-ing is very competitive. You don't necessarily want to kill the person that plays before you or after you, but you do want to show that you've got skills. Men are able to play with each other with a level of toughness, of competitiveness, that doesn't have to be murderous. But the competition between women can get fucked up really quickly. And it's because of this ongoing theme of "I have to be the only one, because I'm exceptional. I'm doing something exceptional, and I'm the queen."

Tricia Rose: And so that identity, that positionality is a by-product of marginality. And it seems to me that one of the only ways to get around that is to produce collectivities of female creativity.

Beth Coleman: But I don't want to have to produce a female collective that has to be put together, because in many of these male collectives people come together because they have shared interests, and not just because they are men.

Tricia Rose: But there is a way that collectivities are produced, they're not always organically formed. Sometimes you look for people, or you ask about people, or you network in certain ways to bring people together. And I'm not suggesting you should be doing this, or that this is your vision, or something. It's just I think that when you combine your other points with this diva issue, particularly those about access—that wonderful phrase you used about sort of being invisible and learning by osmosis—it becomes important to try to produce a network, especially until these networks are more likely to come together more spontaneously.

NOTES

1. Coined by Paul D. Miller, a.k.a. DJ Spooky, illbient is a style of electronic music. It is similar to ambient music in its use of sounds and noises, but unlike ambient, it is less atmospheric and ethereal and more experimental and incorporates grungier, urban soundscapes.

2. Eq is an abbreviation for equalizer, a device that allows one to adjust the emphasis of sound frequencies.

3. Henry Louis Gates, Jr., *The Signifying Monkey* (New York: Oxford University Press, 1988).

Chapter Nine

Black Secret Technology
Detroit Techno and the Information Age

Ben Williams

In 1985 Juan Atkins released the first record on his Metroplex label under the pseudonym Model 500. "No UFOs" is a driving track that plays off the contrast between the warm pulse of funk rhythms and the cold sheen of technological noise. It begins with a simple 4/4 beat made out of a fuzzy scraping noise, then introduces a funky but unmistakably mechanical bass riff, and then eerie, high-pitched sheets of sequencer sounds. Midway through the track a voice chants, "They say there is no go, they say no UFO, why is no head hung high, maybe you'll see them fly," and dissolves into a spacy, vocal dub passage; then the scraping beat returns.

With "No UFOs," Atkins created the blueprint for what came to be known as Detroit techno. The sound did not come out of nowhere; it had its roots in a variety of musical forms. Other earlier releases have a better claim to be the "first techno track," but "No UFOs" marked the first time that all the ingredients that went into techno coalesced into something more than the sum of their parts. It was post-human in affect if not quite in construction, deliberately using only machines and processing the voice heavily on those rare occasions when it was used. It was cinematic, evoking, by turns, gothic scenarios of decaying urban centers and transcendent images of consciousness riding the electronic airwaves. And, alongside such other Atkins tracks of the period as "Infoworld" and "Technicolor," it introduced a self-consciously science-fictional music that predicted an information age that was then just emerging, but is now ubiquitous.[1]

Techno was recognizably related to the presiding utopian—and dystopian—prophecies of its time: Alvin Toffler's book *The Third Wave* (published in 1980), Ridley Scott's movie *Blade Runner* (released in 1982), William Gibson's novel *Neuromancer* (published in 1984). Yet techno's history in America has been a relatively secret one. In some ways, this is unsurprising: a sound that embraces anonymity, alienation, and Europe is out of step with a national musical culture whose myths revolve around identity, authenticity, and homeland. But despite its hybrid roots, techno is a specifically African American variation on the themes of inner-city collapse explored by Toffler, Scott, and Gibson, a variation that teleports African diasporic traditions into the disembodied world of computer networks. And even as techno soundtracks the decaying industry of Detroit, it leaves the city behind for the new global space of postindustrial capitalism.

We Are the Robots: Kraftwerk, Juan Atkins, and the Origins of Techno

As with any musical genre, it is difficult to conclusively pinpoint a single moment as the "beginning" of techno. The first Detroit tracks were preceded by a hybrid history of influences—some more prescient in retrospect than others—split between American and European sources whose common ground lay in their experimentation with the electronic instruments that were just beginning to reach the mass market in the early 1970s. The story begins at a variety of far-flung points: the electronic fusion jazz of Herbie Hancock and Miles Davis introduced abstract machine-generated textures into a polyrhythmic framework; the synthesizer playing of Larry Young and Parliament/Funkadelic's Bernie Worrell layered electronic blocks of sound with little concern for harmony;[2] the sequencer disco of Italy's Giorgio Moroder drew erotic overtones from repetitive rhythms;[3] the sixty-minutes-plus mantra of Manuel Gottsching's fusion guitar and drum machine piece *E2E4* exploited repetition for its ability to produce a sense of timelessness; and early English pop adaptors like the Human League, Depeche Mode, and Simple Minds used synthesizers to create doomy atmospheres of New Europa.

The clearest precedent, however, was obviously set by the German group Kraftwerk, who invented techno's most basic metaphor—the symbiosis of man and machine—in the 1970s. Led by Ralf Hütter and

Florian Schneider, Kraftwerk started out as a rock band called the Organisation in 1971, recording neo-psychedelic music that combined repetitive rhythms and primitive synthesizer effects, before switching to a name that reflected their new, totally mechanical aesthetic. Live, Kraftwerk stood behind banks of synthesizers and drum machines while robot figures performed in their place out front; on record, they honed rhythm to a rigid, motorized pulse that was first realized on the twenty-two-minute title track of their 1974 album *Autobahn*. The record jacket complemented the purity of sound perfectly, depicting cars streaming serenely along a mittel-European freeway that cut precisely through distant mountains.

Kraftwerk first entered African American music in 1982, through the ministrations of the hip hop pioneer Afrika Bambaataa. To Bambaata, the German group sounded like something familiar, but irrevocably different: the precise rhythms of James Brown's funk reconfigured and beamed back to America with self-parodically Teutonic rigidity.[4] Kraftwerk's "Trans-Europe Express" proved to be a dance floor hit in New York; when Bambaataa mixed the 12" single into his DJ set, it would keep people moving long enough for him to go to the bathroom or smoke a cigarette.[5] Thus inspired, he went on to sample the sweeping synthesizers of "Trans-Europe Express" on his epochal *Planet Rock* 12", surrounding them with handclaps, raps, and call and response shouts, and tweaking the flat, metronomic bassline into a fat pulse.

With its oddly mechanical funkiness and spaced-out humor, "Planet Rock" took hip hop one step forward and provided the inspiration for its first spin-off music, electro. If the early 1980s video game culture that sprang up around "Space Invaders" and its ilk provided many kids with their first "learning experience" with computer technology, electro, with its cartoon world of primitive bleeping, bizarre mythology, and secret language, was the unconscious expression of that encounter.[6] Electro would be refined into a music that dealt even more explicitly with electronic technology; Cybotron's 1982 album *Enter* was the most immediate manifestation of that refinement. The production team behind *Enter*, Juan Atkins and Rick Davies, came not from New York but Detroit. Atkins and Davies took the raw good humor of electro, pared the vocals down to minimal slogans (sung by Davies's heavily processed voice in the style of the British New Wave synthpop of the period), and overlaid more complex synthesizer patterns; they also added wailing guitar pyrotechnics on many tracks. But on "Clear," the standout track of the album, there were no guitars, and the result was a bass-driven, stripped-down groove that relied on grating textures and an air of abstract menace to create a futuristic aura.

Davies and Atkins had met in a "future studies" class at Washtenaw Community College in Ypsilanti, which Atkins attended in order to study data processing after reading a Giorgio Moroder album sleeve that described the sequencers the Italian producer had used to create his metronomic disco epics. After realizing he didn't need to be able to program computers to use electronic instruments, Atkins dropped the course, but not before encountering the work of Alvin Toffler.[7] In his book *The Third Wave*, Toffler articulated America's impending transition to a postindustrial, high-tech economy in a distinctly utopian manner; in the process, he also popularized many of the most enduring myths of what is now known as "the new economy." *The Third Wave* outlines such now-familiar concepts as the fragmentation of consensus caused by a vastly expanded media "blip culture"; the increased flexibility of mass production, and the niche marketing that it would enable; the "knowledge workers" who would drive the new information economy; and the suburban "telecommuters" who would abandon the office and the inner city to work at home.[8]

While he tends to be coy on the subject—Atkins has admitted that he was influenced in at least a general sense by Toffler—it would be hardly surprising if he found the futurist inspiring, since he lived in the paradigmatic example of the "Second Wave" industrial economy whose overthrow Toffler described. In the decades following World War II, Detroit's auto industry was the embodiment of a heavily localized market that also dominated globally; the combination of ever-increasing mechanization of production and new forms of global outsourcing, however, had all but eroded that dominance by the 1980s.

"Night Drive (Thru Babylon)": The Global Economy and the Inner City

For Robert J. S. Ross and Kent C. Trachte, as outlined in their study *Global Capitalism: The New Leviathan*, Detroit is nothing less than the epitome of the American transition to full-scale global capitalism that took place in the seventies and eighties.[9] Detroit was at the core of America's industrial manufacturing heartland during its postwar monopoly period; the city's high standard of living and prosperous center, where the heart of its industry was located, embodied the country's success. Led by Ford, GM, and Chrysler, the Michigan region dominated global production of automobiles, electrical machinery, steel, and rubber. As a consequence of the stable economic conditions created by this monopoly—which included fixed prices and powerful unions—workers experienced unprecedented salary and employment levels.

In the late 1960s, however, came the beginning of the end of this golden era: profits declined throughout the industry, and the foreign import share of the U.S. auto market doubled. As U.S. domination of the global auto industry waned, so price competition returned and market shares became much less stable. In order to maintain profit levels in the face of competition with foreign manufacturers, U.S. companies had to control wages; but the strength and concentration of unions in the Detroit region meant that the car companies needed a new bargaining lever in order to enforce pay cuts.

A combination of substitution of machines for labor and global outsourcing—the location of production in low-wage and/or high-subsidy areas—proved to be the answer. In the 1970s, the auto manufacturers began investing substantially in new plants and equipment abroad, resulting in a declining share of worldwide production in the Detroit area. The most marked increase in foreign investment occurred between the end of 1978 and the first half of 1982, the height of the U.S. auto crisis and a period that Ross and Trachte see as marking the shift from monopoly to global capitalism.

This new mobility of capital allowed the auto manufacturers to force major concessions from the city council, the unions, and the welfare state. In 1982 the United Auto Workers gave previously won benefits back to Ford and GM in order to compete with foreign labor. And while those who had jobs still enjoyed monopoly-era wages, their numbers were declining drastically: from 1978 to 1982, nearly 270,000 production jobs were lost, and unemployment in the motor vehicle industry averaged 17 percent. In 1980 the unemployment level among Detroit's majority black population was 27 percent. The net effect of the restructuring was that Detroit's municipal government faced a fiscal crisis, resulting in drastic expenditure reduction in human services, public services, and economic development at a time when the need for these services was at a postwar high. In January 1983 Mayor Coleman Young declared a state of emergency.

While Detroit's techno producers would make music that evoked, mourned, and sometimes mythologized these inner-city problems, the cultural milieu out of which techno emerged was suburban and affluent. In his book *Techno Rebels,* Detroit native Dan Sicko details a club scene divided geographically by Woodward Avenue, a twenty-seven-mile stretch of road from Detroit to the northern suburb of Pontiac. On the east side were the less affluent "jits," on the west were the upper-middle-class "preps." According to Sicko, the party scene in the early 1980s was oriented around "preps" and their explicitly exclusionary high school social clubs; entrepreneurial-minded promoters produced highly successful parties that often specified, "no hats and no canes"—code for "no jits and no thugs."[10] The kids were into *American Gigolo, GQ,* designer labels, and Italian disco, which became the staple sound of the circuit.[11] The clubs had names like "Chari Vari" and "Snobs" and the track cited by many as the "first" example of Detroit techno was actually named "Sharevari." The lyrics that accompany its Moroder-esque groove reference such "European" delicacies as white wine and cheese. In late 1982 and early 1983, however, some of these barriers weakened a little, at least temporarily, as preps started to drive over to the looser, less organized, and more inclusive east side parties, which were also the early home of electro in Detroit.

Juan Atkins's family lived on the west side, thirty miles away from the collapsing city center, in the mostly white suburb of Belleville.[12] As a Belleville teenager in the late seventies, Atkins met the two other producers who would provide the initial burst of invention behind techno, Derrick May and Kevin Saunderson. He teamed up with May for occasional DJ gigs on the party circuit under the moniker Deep Space, though the pair was never a major force. They found their influences on the airwaves: the legendary local DJ the Electrifying Mojo broadcast a radio show that mixed up New Wave synth pop, hip hop, P-Funk, electro, early ambient music, Kraftwerk, and sci-fi movie soundtracks—everything, in fact, that would be condensed into techno.

Though Detroit was in many ways an archetypal postindustrial city by the early 1980s—an empty center, surrounded by a metropolitan area composed of multiple suburbs—what marked it out from, say, Los Angeles, was the unavoidable presence of the industrial past in the form of the bombed-out (or occasionally repurposed) shells of the great automobile factories. "The fabulous ruins of Detroit" (as a Web site devoted to documenting them is titled) litter the city, particularly its downtown area: giant, imposing structures like the once-thriving Michigan Central Station and the J. L. Hudson department store can be seen from a "People Mover" train, the first component of a never-completed mass transit system built in the 1970s.[13] Combined with suburbanization, the lack of a vital downtown and public transportation produced a city of relatively isolated, self-enclosed communities connected only by the automobile.

Perhaps aware of these contradictions, Atkins chose to locate his Metroplex headquarters in the city center, in a building in Detroit's Eastern Market; May's Transmat label and Saunderson's KMS quickly joined him there. This area ("Techno Boulevard") became the center of the scene for years to come, as the producers who followed in the footsteps of May, Atkins, and Saunderson got their starts working for these labels while their owners were abroad, reaping the fruits of European success. In combining high-tech minimalism with a palpable aura of melancholy, Detroit techno soundtracked this environment and the faded dreams it represented. The music was a response to the painful contradictions of the city's changing economy, a form of science fiction that provided an aesthetic solution to Detroit's problems. Escapist in a positive rather than negative sense, techno provided a way for its producers to both symbolically and literally imagine themselves into the future. They gave their labels names that sounded like *Bladerunner*-style futuristic corporations; decorated their record sleeves (when there was any decoration at all) with a metallic sheen and anonymous, multinational-style logos; and disguised themselves behind a host of brand-name pseudonyms—Channel One, Tronik House, Suburban Knight, Drexciya.

Juan Atkins created sonic worlds that soundtracked the collapsing inner city ("Night Drive [Thru Babylon]"), outer space ("No UFOs"), and the info-tech future ("Infoworld"), worlds that made the future sound like a cold, grim place to live. Yet he also seemed intoxicated by, and accepting of, the powerful technological forces that would presumably dominate it. As he famously stated, "Berry Gordy built the Motown sound on the same principles as the conveyor belt system of Ford. . . . Today the plants don't work that way. They use computers and robots to build the cars. . . . I'm probably more interested in Ford's robots than Berry Gordy's music."[14] Though Atkins would later qualify this quote by saying that he intended no disrespect to Gordy, it nevertheless articulates the changes Atkins would make to African American musical tradition. (Gordy, after all, explicitly followed an industrial model for the running of Motown; his "conveyor belt" system was intended to produce finely tuned musical products educated not only in sound, but in all aspects of comportment, demeanor, and presentation. Atkins instinctively homed in on the robots that replaced humans on the conveyor belts and connected them to the decline of the industrial model, both economically and musically.)

He was not the first to make this connection. As Tricia Rose puts it, Bambaata and other rappers saw, in Kraftwerk, "an understanding of themselves as already having been robots."[15] Becoming robots was, for African American musicians, a subliminally political act, the ramifications of which can be read as both a form of self-empowerment and an identification with otherness, whether technological or racial. The historical recognition of oneself as a robot—or, following Rose, a unit of capitalist labor—is accompanied by the simultaneous inversion of that identity so that it becomes a form of power; robots, after all, are stronger, tougher, and more enduring than mere humans.

Rather than criticizing the deteriorating conditions of Detroit, techno producers took those conditions as a fatalistic given to build from, a real-life sci-fi scenario. As the producer Stacey Pullen, who recorded a number of albums for Derrick May's Transmat label in the early 1990s, puts it,

> I grew up in an era when it was an all black neighborhood. Detroit is an 80% black city and the race riots of 1967 left it in shambles and neglect. A lot of businesses left. So we grew up as a product of our environment. There was a dark cloud over Detroit. . . . That's why I think we portray a futuristic way of thinking, and are influenced by science fiction movies. If you watch a typical science fiction movie, it's about a city that has been blown up or neglected. The government is corrupt and basically that explains it in a nutshell.[16]

Faking the Funk: Derrick May, Kevin Saunderson, and African American Tradition

But if the sound embraced the future, it also recast it by applying techniques derived from African American musical tradition to technology. Aside from Motown, another immediate regional influence was the compelling fusion of high-tech rhythm and space age fantasy created by the Detroit native George Clinton. Techno expanded on Clinton's aesthetic by reconfiguring its defining qualities—relentlessly precise rhythm loops, driven by basslines that bleed around the edges of their notes, and disruptive noise—to fit synthesizer and drum machine technology. The Roland drum machine supplied perfectly syncopated rhythm patterns; tracks like Kevin Saunderson's "Just Another Chance" featured deep, ruptured basslines at the front of the mix and substituted the atonal shrieks of James Brown and the jazz solos of Maceo Parker with bleeps, squeaks, and scrapes of technology overloaded, interrupted, and twisted.[17]

Techno took those experiments in dissonance further. Eschewing samples of live instruments in favor of raw, mostly atonal machine noises, techno's vocabulary involved a heightened emphasis on texture and rhythm and an approach to sound elements as building blocks. This vocabulary included such elements as the abrupt fading in and out of sounds, the increasing and decreasing of their intensities, and the use of varieties of graininess. The result was a shifting interplay of sound patterns that were not, however, simply abstract: for all its conceptual tendencies, techno is first and foremost dance music, experienced most immediately as an endlessly propulsive rhythm and almost always subject to the rigid organization that implies. Although basic piano motifs, diva vocals, and even occasionally saxophones appeared in various techno tracks, the music subordinated almost all its elements to rhythmic functions; the exception being its blocks of synthesized sound, which functioned more as atmospheric texture than melody. This aspect of techno is illustrated best by the work of Derrick May, the most celebrated producer of techno's late-eighties period. Working under the names Rhythim Is Rhythim, Mayday, and R-Tyme, May produced a series of seminal tracks from 1987 to 1990 that combined unparalleled rhythmic sophistication with emotional use of sweeping synths, updating elements of both funk and disco.

Alongside the work of Kevin Saunderson, whose most populist

tracks featured giant melodic hooks, May tracks like "Strings of Life" conveyed an anthemic quality that was close to another of techno's main influences, the house music of Chicago (itself derived from disco). In fact, the distinction between techno and house has always been blurred at the edges; though the latter tends to be distinguished by its vocals and uplifting piano melodies, the instrumental strain of house has often dispensed with both in the pursuit of pure, repetitive textures.

Despite all these influences, techno also significantly departed from African diasporic music traditions, which may have prevented its acceptance in centers of black music like New York. Where hip hop firmly inscribed its blackness within its machine-created grooves by recontextualizing the oral tradition and sampling canonical funk grooves, techno's more total embrace of a mechanized sound and aesthetic moved away from such traditional signifiers of the black body as gospel-derived vocal harmonies, oral narrative, and instrumental virtuosity.[18] Hip hop had its black street audience; house had its black gay audience; techno had almost no black audience outside Detroit, and would achieve far greater recognition in Europe. After their early work became popular in Britain, May, Saunderson, and Atkins left Detroit for extended periods to DJ abroad.

The fact that techno embraced, rather than hid, the irreducible element of mediation in its sound may well have reduced its appeal to a larger African American audience in search of recognizable people to identify with.[19] Where hip hop tended to use recognizable samples that were intended to trigger emotional memories/histories, techno treated its found sounds as interchangeable elements of an endless data stream. Techno lets the machines take over to a limited degree: it is impossible to imagine a human being physically playing this music in real time. Even a DJ is one step removed from the music by virtue of the fact that he manipulates flows of sound, as opposed to physically producing it.

The Stars My Destination: Underground Resistance and Cyborg Subjectivity

This tension between African American tradition and technology that seemed to produce music devoid of cultural and human reference points was addressed most explicitly by the group Underground Resistance (UR), who picked up where May, Atkins, and Saunderson left off in the early 1990s. If the initial Detroit producers were ambivalently aware of techno's more philosophical implications, this shifting collective was militantly self-conscious, adopting the image and rhetoric of guerrilla warfare in order to combat economic exploitation by major record labels. But while they preached a localized self-sufficiency that was in a sense the opposite of the early techno producers' quick flight to Europe, UR also had a mystical side that drew on American Indian and African American mythology. The combination would result in a new form of racial subjectivity.

The core members were Mad Mike Banks and Jeff Mills.[20] Before Underground Resistance's first release in 1991, Mills had held down a regular DJ gig at WNET and had been a member of the local industrial music group Final Cut; Banks had played bass in one of George Clinton's many spin-off bands, the Brides of Funkenstein. As Public Enemy had infused rap with a social/political consciousness that imbued the music as a whole with a rebellious significance that kids worldwide could identify with, so Banks and Mills supplied techno with similarly appealing manifestos of resistance through media warfare. The front cover of the first Underground Resistance album, 1992's *Revolution for Change*, depicts shadowy military-looking figures caught in the grainy blue textures of a surveillance camera; the back has a picture of Detroit in flames with the logo "Hard Music from a City" emblazoned across it. Next to the picture, the manifesto begins:

> Underground Resistance is a label for a movement. A movement that wants change by Sonic Revolution. We urge you to join the Resistance and help us to combat the mediocre audio and visual programming that is being fed to the inhabitants of earth, this programming is stagnating the minds of the people; building a wall between races and preventing world peace. It is this wall we are going to smash.

The music inside the sleeve was raw, abrasive, and pummeling: Underground Resistance soundtracked the body under assault, attacked by radioactive forces, tracked by high-tech surveillance equipment, and armored against invasion. "Riot" opens with the sound of a massed crowd chanting the title over a metronome beat; "Eliminator" and "Adrenaline" were similarly aggressive, favoring martial, unforgiving beats that would prove to have particular influence on Europe's rave scene.

Underground Resistance's sound conjured up a robot stalking the postindustrial streets, flowing with the relentless pace and precision of machine time: the protagonists in the films *Robocop* (set in Detroit) and *Terminator*. The military aspect of their manifesto echoes conspiracy theories about corporate/governmental mind control of the black body through electrical signals. This aural warfare could be considered a response to a secret history of African American oppression by technology. As Mark Dery has pointed out in his essay "Black to the Future,"

> African Americans, in a very real sense, are the descendants of alien abductees; they inhabit a sci-fi nightmare in which unseen but no less impassable force fields of intolerance frustrate their movements; official histories undo what has been done; and technology is too often brought to bear on black bodies (branding, forced sterilization, the Tuskegee experiment, and tasers come readily to mind).[21]

But the militarism provoked by this treatment of African Americans is accompanied by a transcendental vision that aurally imagined an extraterrestrial human existence. But full frontal assault was only one weapon in the UR armory—and, in light of their almost total avoidance of visibility and publicity, the least valued. Elsewhere on *Revolution for Change*, deep, dark tracks like "Quadrasonic" and "Predator" conducted subtle tonal modulations over mantra-like beats. These recordings sounded like guerrillas on an assault mission deep in enemy territory, communicating by sonar devices only. The second half of the text on the *Revolution for Change* album reads,

> By using the untapped energy potential of sound we are going to destroy this wall much the same as certain frequencies shatter glass. Techno is a music based in experimentation; it is sacred to no one race; it has no definitive sound. It is music for the future of the Human Race. . . . By simply communicating through sound, Techno has brought people of all different nationalities together under one roof to enjoy themselves. Isn't it obvious that music and dance are the keys to the universe? So called primitive animals and tribal humans have known this for thousands of years! We urge all brothers and sisters of the under-

> ground to create and transmit their tones and frequencies no matter how so called primitive their equipment may be. Transmit these tones and wreak havoc on the programmers!

The keynote here is rapt awe: an astral aesthetic that often owes as much to the cosmic jazz of the early seventies as it does to Kraftwerk and P-Funk. The height of Underground Resistance's astral ambitions can be found on the *Galaxy to Galaxy* album and the Red Planet series of releases.[22] These tracks mix rubbery basslines with oscillating sequencer patterns and spiraling 303 lines that, in their blissful textural caresses, express a certain cosmic rapture.[23] In this context, Tricia Rose's association of the feeling of "awe" with the mastery of technology takes on a rather different significance.[24] Mastery of the machines shades into transcendence of and through them; a certain analogy of freedom paradoxically arises from the rigidity of techno's relentless beats. The substitution of the beginnings and endings of chord development with endless repetition focuses attention on micro-textural shifts, allowing them (at best) to produce emotional effects that create a form of technological sublime.

It is tempting to read this music as an invitation to escape this world, and with it racialized forms of oppression; as an imaginary dissolving of the body amidst the pure abstraction of technological special effects. But we should remember that for Juan Atkins, there were no UFOs. Rather than reading these metaphors of space travel and intergalactic communication as an imaginary ride in the mothership, we may also analyze them as allegorizing bodies in motion, finding ways to accommodate themselves to the ever-shifting technologies of capitalist innovation. In this reading, rather than erasing the body, Detroit techno's astral strain regrounds it within the global electromagnetic sphere of postindustrial communications technologies.

The obsessive focus of techno on purely mechanical forms of sound organization—which exist nowhere in the "real" world of real-time instrumentation, but only in the parallel universe of the studio—is a musical equivalent to the special effects of sci-fi movies and the lyric language of literary science fiction.[25] Outer space is the natural figure for the expression of this reality: the radically alienated environment it represents and the weightlessness experienced by the astronaut who inhabits it constitute a world in which human experience is necessarily mediated by technology, and cast adrift from natural reference points.

Hence the outer space fantasies of Underground Resistance emphasize the most abstract of machine noises and textures to create a more fluid and less mechanical sound than on their more militant recordings.

The body that this music creates is not armored against the world with robot technology, but rather open and redefined by technology. Atkins's robot has been replaced by the cyborg, and the difference, as Claudia Springer points out, is crucial: "While robots represent the acclaim and fear evoked by industrial age machines for their ability to function independently of humans, cyborgs incorporate rather than exclude humans, and in so doing erase the distinctions previously assumed to distinguish humanity from technology."[26] Mike Banks has made this connection himself:

> I love the 303 though I try and think of creative ways of using it, like hi-tech jazz. I always like to include it, just like a musician in my band. All my instruments I've got a personal electronic relationship with. 'Cos I have electricity in my body too. When we touch they recognize me and I recognize them. I think *Terminator* is definitely what's happening man—machines have emotions. They can become intelligent and they will one day.[27]

The astral recordings of Underground Resistance constitute a representation of that zone as lyrical as the language William Gibson used to conceptualize cyberspace.[28] UR took Juan Atkins's fascination with Toffler's *Third Wave* one step further, to the point where the local city connects with the new electronic space of global capitalism. Detroit is imagined not as a crippled Second-Wave industrial city, but as an alternate Silicon Valley in which the robots have downed their tools at the conveyor belts, taken to the datawaves, and begun transmitting communiqués of subversion worldwide.

The Journey Home (Future): Drexciya and Cyber-Africa

The mapping of these new global spaces is not, however, limited to outer space metaphors; indeed, it finds its most explicit representation rather closer to home: beneath the ocean, whose watery embrace supplies a similarly fluid context. It is in the work of the Underground Resistance side-project Drexciya (a group composed of two producers who remain anonymous) that the aquatic metaphor is developed most fully, in terms of both sound and concept.

Rather than the raptures of cosmic jazz, Drexciya turned back toward the willfully clunky sounds of electro in order to create a suitably squelchy atmosphere for their amphibious mythologies. Speculating about a race of "Mutant Gillmen (An Experiment Gone Wrong)," who surfed the "Aquabahn" to a "Bubble Metropolis," Drexciya pursued a raw sound reminiscent of Model 500, but replaced the futuristic sheen of Juan Atkins's work with sequencer patterns that sound like bubbles rising to the surface of the ocean, heavily synthesized vocal soundbites, and peculiarly tactile textures (memorably described by Kodwo Eshun as "acrid frequencies [that] clench the nerves like tazers").[29]

The sleeve notes of the compilation CD *The Quest*, released as the first and supposedly final Drexciya album after a long series of 12", feature a map divided into four stages:

1. *The Slave Trade (1655–1867)*—a triangular route between Africa, southern Europe, and the Americas.
2. *Migration Route of Rural Blacks to Northern Cities (1930s–1940s)*—a large circle radiating from the South with arrows pointing West, North, and Northeast.
3. *Techno leaves Detroit, Spreads worldwide (1988)*—arrows pointing in every direction from Michigan.
4. *The Journey Home (Future)*—multiple routes pointing from North and South America to Africa.

The final image of techno's infiltration of global communication circuits that accompanies "the journey home" constitutes the logical conclusion of the program first outlined on Underground Resistance's debut album notes. The rest of *The Quest*'s sleeve notes, which are devoted to speculation about the existence of a race of sea creatures mutated from pregnant African slaves who were thrown overboard during the passage to America, supply a corresponding subjectivity in the form of an origin myth:

> Are Drexciyans water-breathing, aquatically-mutated descendants of those unfortunate victims of human greed? Have they been spared by God to teach us or terrorize us? Did they migrate from the Gulf of Mexico to the Mississippi river basin and to the great lakes of Michigan? Do they walk among us? Are they more advanced than us? How and why do they make their strange music? What is their quest?

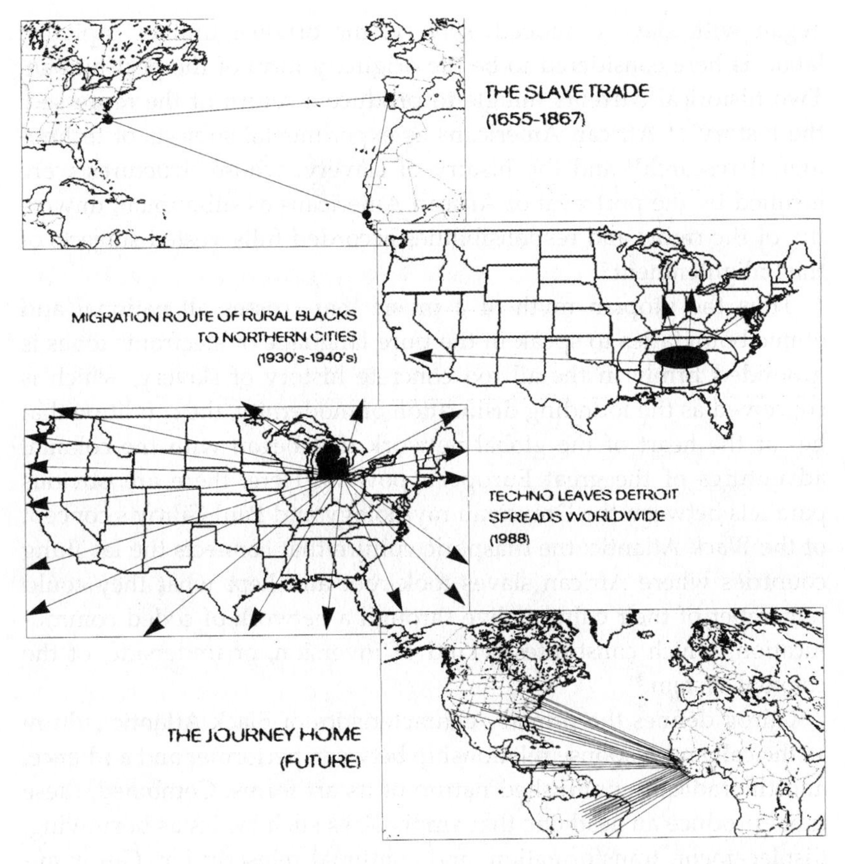

Liner notes from *The Quest* CD by Drexciya. Mad Mike Banks/Submerge Records.

The theme of mutated beings is continued on the most recent album on the Underground Resistance label, 1998's *Interstellar Fugitives*, which speculates on the existence of a mutant strain of African American genes that has produced such historical "warrior" figures as Martin Luther King and Nelson Mandela. What these mutant beings have in common with the cyborg, rather than the robot, is their theorization of a fundamental technological alteration in, rather than extension of, what it means to be human.

In this context, the mechanical metaphors we have been tracking extend beyond signifying post-humanity to embody a history that began with slavery; indeed, slavery, the original unit of capitalist labor, is here considered to be the originary form of the post-human. Two historical currents mingle to produce a return of the repressed: the history of African Americans as experimental subjects of technological research[30] and the history of slavery, whose structures were justified by the portrayal of African Americans as subhuman, unworthy of the rights and responsibilities accorded fully vested subjects of the Enlightenment.[31]

Thus the utopian myth of a music that crosses all national and ethnic boundaries to speak in the pure language of electronic tones is grounded firmly in the all too concrete history of slavery, which is conceived as the founding dislocation of modernity, the condition that lies at the heart of the global network that began with the colonial adventures of the great European powers. Here, there are obvious parallels between the Drexciyan mythology and Paul Gilroy's concept of the Black Atlantic: the diasporic culture that connects the far-flung countries where African slaves took root and kept what they could remember of their culture alive through a network of coded communication, which constitutes a kind of inversion, or underside, of the colonial system.[32]

Gilroy defines the primary characteristics of Black Atlantic culture as the call and response relationship between performer and audience, and the radically unfinished nature of its art forms. Combined, these traits produce an aesthetic that emphasizes such tactics as borrowing, displacement, transformation, and continual reinscription. Gilroy analyzes the remix as a form of call and response, seeing its deconstruction of an original song as demanding supplementary creative input from either the DJ or the producer, and thus enabling a passage from aesthetic object to collective event.

But he refuses to reify these qualities into a pluralist, fragmentary opposition to all that is fixed, stable, and complete, observing that they produce forms of identity, such as "the imaginary effect of an internal racial core or essence," out of the identification between performer and crowd.[33] The dislocation of slavery produces the identity of alienation, an ever-changing same embodied most brilliantly in Black Atlantic musical cultures. But where Gilroy concludes that "The most important lesson music still has to teach us is that its inner secrets and its ethnic rules can be taught and learned," his analysis continues only as far as hip hop, and ends by chiding it for "betraying" its own radicalism and falling back on a rhetoric of racial essence.[34] Drexciya's cyborg myths constitute an attempt to imagine a kind of subjectivity that can live with these paradoxes. They imagine an origin of a blackness that, since it is characterized by the concept of mutation rather than that of skin color, cannot be reduced only to blackness.

Similarly, Detroit techno represents an African American aesthetic approach to technology, but one that has been so muddied by cultural exchange that it cannot be tied to a single physical place. Global dissemination and mongrelization of music have occurred since European colonialism began "discovering" the New World; the difference today is merely one of degree, of the speed with which records are exported around the world or data files are downloaded off the Web, and of the ease with which they can be copied and incorporated into local styles.

That matter of degree, however, minimizes the element of existential life experience involved in the creation of music in favor of accelerated transmission of cultural structures via electronic communication. Techno replaces the blues philosophies of suffering that have underpinned twentieth-century African American music with an aesthetic of reprogramming. "Blackness" becomes production technique; alienation is converted into cyborg identity; and the practice of international musical data exchange becomes a utopian myth of nonproperty-based, "open source" collaboration that functions as a resolution to the contradictions and inequities of global electronic capitalism. Drexciya's project of mapping the Black Atlantis can be considered an allegory for this "ocean of sound," a journey home that leads to the future.

What is this future? It is postapocalyptic, but less in the literal sense of a post–nuclear war narrative, and more in the metaphorical sense of being built on the ruins of industrial culture, out of electronic technologies that seem to offer the possibility of dissolving the mass media of the age and returning to preindustrial, "gift economy" forms of communication that offer more localized forms of response.

Yet there are ironies here. It would be a mistake to read this future as a simple return to the past, to "primitive," nonlinear forms of consciousness. New forms of technology are supposed to heal the excesses of the old ones; these talking drums communicate within an electronic environment far removed from the rain forest or the savannah. The physical landscape is more feudal than tribal: a post–welfare state economy where the market rules and global corporations produce economic action-at-a-distance effects on local lives. The city itself has passed beyond the sensory comprehension of the individual. It may even be that sound, free of the objectifying tendencies that accompany visual representation, is best equipped to map—certainly, to evoke—these new spaces. But the particular kind of communications space that techno inhabits and mythologizes is only imaginable in an American economy that has, for the most part, moved its industry to other parts of the world, where labor is cheap and laws less stringent, in order to shift toward media and entertainment modes of production. In this sense, Detroit techno's global utopia remains firmly local. The final destination of cyber-Africa has yet to be imagined.

NOTES

1. Considered in this predictive capacity, techno seems to bolster Jacques Attali's thesis that music has the capacity to act as a herald of social transformation. Music, he says, is prophetic: a "way of perceiving the world" that "makes mutations audible" and can embody emerging social and economic formations in aesthetic terms. Jacques Attali, *Noise: The Political Economy of Music* (Manchester: Manchester University Press, 1985), 4.

2. Larry Young developed a Coltrane-influenced modal style on the Hammond organ in the mid-1960s, before serving time with all the major fusion band leaders of the 1970s—Miles Davis, John McLaughlin, and Tony Williams.

3. Most obviously on Donna Summer's "I Feel Love."

4. Kraftwerk were self-confessed fans of Motown and James Brown. As member Karl Bartos put it, "We always tried to make an American rhythm feel, with a European approach to harmony and melody." Quoted in Dan Sicko, *Techno Rebels* (New York: Billboard Books, 1999), 26.

5. David Toop, *Rap Attack 2: African Rap to Global Hip Hop* (New York: Serpent's Tail, 1991), 130.

6. Ibid., 146.

7. Dan Sicko, "The Roots of Techno," *Wired* 2, no. 7 (July 1994).

8. Alvin Toffler, *The Third Wave* (New York: Bantam, 1980).

9. The following material is drawn from Robert J. S. Ross and Kent C. Trachte, *Global Capitalism: The New Leviathan* (Albany: State University of New York Press, 1990), 115–46. For more on Detroit and postindustrialism, see

Thomas J. Sugrue, *The Origins of the Urban Crisis: Race and Inequality in Postwar Detroit* (Princeton: Princeton University Press, 1996).

10. Sicko, *Techno Rebels*, 39–40.

11. The pioneering techno DJ Jeff Mills has described the film *American Gigolo* as the essence of techno. See Simon Reynolds, *Generation Ecstasy: Into the World of Techno and Rave Culture* (New York: Little, Brown, 1998), 15.

12. The overt identification of Detroit's black techno producers with Europe in general and the very "white" aesthetic of Kraftwerk in particular has been much discussed, most notably by Simon Reynolds and Kodwo Eshun. For Reynolds, Detroit's techno musicians were elitist snobs whose music represented a middle-class "art" aesthetic that emphasized melody and texture over the supposedly working-class musical values of bass and rhythm. While there is a grain of truth to this analysis, it is also heavily overdetermined, perhaps by anxieties of influence, and its material grounding is compromised by the elevation of discrete musical qualities into transcendent markers of class. In the process, some familiar stereotypes are reinforced: middle-class equals mind, working-class equals body (with male/female and white/black dichotomies lurking just behind).

Eshun, on the other hand, reverses this equation with its familiar histories of European/American musical influence—and, not incidentally, authenticity—by asserting that "Kraftwerk are to Techno what Muddy Waters is to the Rolling Stones: the authentic, the origin, the real." For Eshun, this was a move consciously intended to alienate techno and its creators from both America and fixed concepts of blackness and authenticity; it is part of a larger, transnational black musical history that replaces representation with "electronic fictions" that create escape routes from consensus reality by "reprogramming" consciousness. This, as we shall see, has much in common with the Detroit techno collective Underground Resistance; but it also overemphasizes—if not fetishizes—"otherness" (in every sense) while ignoring the economic and historical factors that played into techno's aesthetic. Reynolds, *Generation Ecstasy;* and Kodwo Eshun, *More Brilliant Than the Sun: Adventures in Sonic Fiction* (London: Quartet Books, 1998), 100.

13. Reynolds, *Generation Ecstasy*, 14. The Fabulous Ruins of Detroit site, created and maintained by the local artist Lowell Boileau, can be found at www.bhere.com/ruins/home.htm.

14. Matthew Collin, *Altered State* (New York: Serpent's Tail, 1997), 23.

15. Mark Dery, "Black to the Future: Interviews with Samuel R. Delany, Greg Tate, and Tricia Rose," in *Flame Wars: The Discourse of Cyberculture*, ed. Mark Dery (Durham: Duke University Press, 1994), 213–14. Emphasis in original.

16. "Stacey Pullen Interview with Joanne Wain," *Jockey Slut*, December–January 1996–97.

17. Hip hop, of course, had already come up with its own ways of juicing the funk blueprint with technology, and the more sonically adventurous strains that developed in the late eighties in particular have much in common with techno in terms of their approach to sound construction. Indeed techno, with its evocation of a Tofflerian world of edge city info-commuters, could be considered to constitute a "suburban" complement to what Tricia Rose described as hip hop's response by "urban people of color to the postindustrial landscape." Dery, "Black to the Future," 213. For Rose, hip hop is as much a sonic aesthetic as a verbal one, characterized by such qualities as leakage, multilayering, low frequencies, rupture, openness, and forces in motion. Tricia Rose, *Black Noise: Rap Music and Black Culture in Contemporary America* (Hanover: Wesleyan University Press, 1994), 74–80.

18. Although some producers, most notably Blake Baxter, did include recognizably human vocals, even these were treated in a way that produced a certain distanced alienation.

19. Though not to a British music industry perpetually in search of new sounds to hype.

20. Others in the long list of producers who converged around UR and Submerge (www.submerge.com), the umbrella organization of labels, producers, and businesses of which it is a part, at some point in their careers include Robert Hood, Suburban Knight (a.k.a. James Pennington), and Drexciya.

21. Dery, "Black to the Future," 180.

22. On *Galaxy to Galaxy*, tracks like "Hi-Tech Jazz" score the cheery melodies of an oldtime swing band for machines, while "Star Sailing" and "Deep Space 9" balance atonal noise, stripped-down rhythms, and simple melody lines. The Red Planet series, credited for many years to "The Martian" (a.k.a. Will Thomas), began with a sample of the Black Panther Party's slogan, "The spirit of the people is greater than the man's technology" on the track "Last Transmission from Earth," and later recordings like "Stardancer," "Ghostdancer," "Windwalker," "Sky Painter," "Medicine Man," and "War Dance" evoke a kind of mystical warrior relationship to machines. These titles are notable because Mike Banks's mother is Blackfoot Indian and UR tracks explicitly reference this heritage. For more, see Reynolds, *Generation Ecstasy*, 232.

23. The title of one track, "Sex in Zero Gravity," probably sums up the desired effect.

24. Dery, "Black to the Future," 212.

25. Science fiction has always drawn on heightened rhetorical forms in order to convey a sense of wonder at the sheer otherness of the future worlds it creates: in literature, this takes the form of lyric language; in film, special effects. Following Samuel Delany, I want to suggest that these forms of rhetoric actually allegorize technology, inasmuch as the "other" spaces that they map are always technologically constituted. In effect, science fictional rhetoric

defamiliarizes the mechanical environment by describing it with heightened and exaggerated forms of language, and thereby foregrounds it. See Scott Bukatman, *Terminal Identity* (Durham: Duke University Press, 1993), 157.

26. Claudia Springer, "The Pleasure of the Interface," *Screen* 32 (autumn 1991): 3.

27. These sleeve notes are credited to "The Unknown Writer." See also Paul Benney, "Nobody Buys Me," *Jockey Slut*, July 1994. Accompanying this transition from robot to cyborg and from natural to technological sublime, according to Bukatman, is a corresponding "shift in the experience and definition of the city from centralized place to 'dispersed non-space.' " The city, he continues, "has passed beyond the sensory powers of the individual" and generated "a related set of representational systems that write the passage of the city beyond the parameters of perception itself." It may even be true that sound, free of the objectifying tendencies that accompany visual representation, is best equipped to map these new spaces that extend beyond direct human comprehension. Bukatman, *Terminal Identity*, 168.

28. Gibson coined the term "cyberspace" in his influential cyberpunk novel *Neuromancer*.

29. Eshun, *More Brilliant Than the Sun*, 83.

30. Research that, as in the comic book mythologies that Underground Resistance drew on, produces beings endowed with extrahuman powers.

31. See also Mark Sinker, "Black Science Fiction," *Wire* 96 (February 1992).

32. See also Kodwo Eshun, "Drexciya," *Wire* 167 (January 1998).

33. Such identifications, furthermore, are not limited by physical constraints, or even to particular types of call and response situations: "This reciprocal relationship can serve as an ideal communicative situation even when the original makers of the music and its eventual consumers are separated in space and time or divided by the technologies of sound reproduction and the commodity form which their art has sought to resist." Paul Gilroy, *The Black Atlantic: Modernity and Double Consciousness* (Cambridge: Harvard University Press, 1993), 102.

34. Ibid., 109.

DISCOGRAPHY

Bambaataa Afrika, and the Soul Sonic Force. *Planet Rock*. 12". Tommy Boy, 1982.

Channel One. Technicolor. 12". Metroplex, 1986.

Cybotron. *Alleys of Your Mind*. 12". Deep Space, 1981.

———. *Enter* (includes "Clear"). LP, Fantasy, 1983.

Drexciya, *The Quest*. CD, Submerge, 1997.

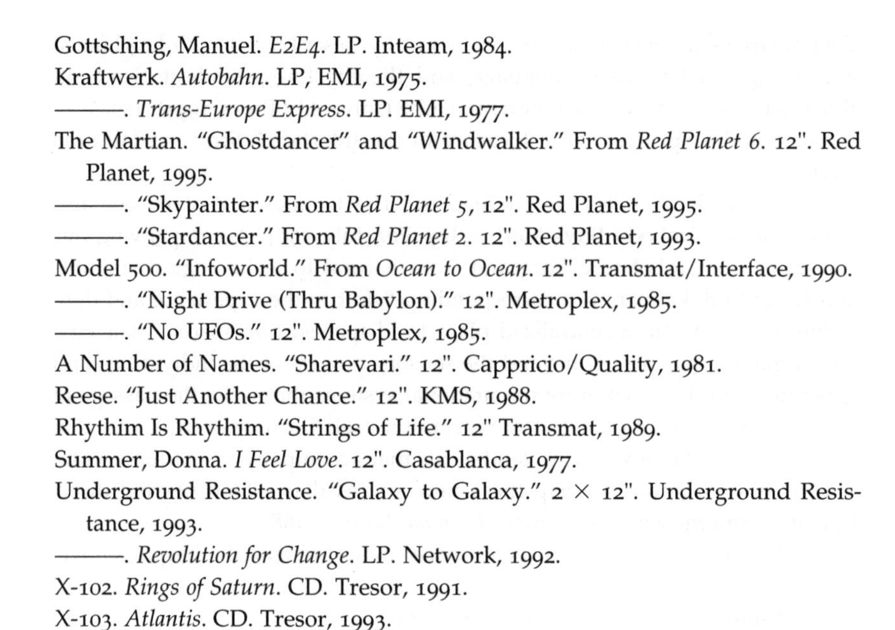

Gottsching, Manuel. *E2E4*. LP. Inteam, 1984.

Kraftwerk. *Autobahn*. LP, EMI, 1975.

———. *Trans-Europe Express*. LP. EMI, 1977.

The Martian. "Ghostdancer" and "Windwalker." From *Red Planet 6*. 12". Red Planet, 1995.

———. "Skypainter." From *Red Planet 5*, 12". Red Planet, 1995.

———. "Stardancer." From *Red Planet 2*. 12". Red Planet, 1993.

Model 500. "Infoworld." From *Ocean to Ocean*. 12". Transmat/Interface, 1990.

———. "Night Drive (Thru Babylon)." 12". Metroplex, 1985.

———. "No UFOs." 12". Metroplex, 1985.

A Number of Names. "Sharevari." 12". Cappricio/Quality, 1981.

Reese. "Just Another Chance." 12". KMS, 1988.

Rhythim Is Rhythim. "Strings of Life." 12" Transmat, 1989.

Summer, Donna. *I Feel Love*. 12". Casablanca, 1977.

Underground Resistance. "Galaxy to Galaxy." 2 × 12". Underground Resistance, 1993.

———. *Revolution for Change*. LP. Network, 1992.

X-102. *Rings of Saturn*. CD. Tresor, 1991.

X-103. *Atlantis*. CD. Tresor, 1993.

Chapter Ten

Tales of an Asiatic Geek Girl

Slant *from Paper to Pixels*

Mimi Nguyen

Okay, it's true, I admit: I used to have a *thing* for Xerox copiers. For a girl obsessed with 'zine-ing, it was, after all, only natural that I develop an equally fervent love-object relationship to the machine, my big dumb prosthesis. It was a long-term affair: I was fifteen when it began, a shiny new punk rocker decked out in riot grrrl gear, armed with equal parts CRASS, Bikini Kill, and Noam Chomsky. I read the fanzine reviews in *MaximumRocknRoll*, sent away for titles like *Assault (With Intent to Free)*, *Unmentionable*, and *Blowin' Chunx* and was subsequently amazed that these wayward kids who were *just like me* (I'd qualify that now, of course) nurtured secret identities as savvy guerrilla media producers, ripping off printers and postage for higher (punk) purposes. Ridiculously accessible in practice, *my* first 'zine was a mostly handwritten thing, rife with stolen graphics, the illicit product of late nights at my mother's work, where I would stage a temporary takeover of the copier room. It was an adventure that made me feel powerful, diffuse, and positively giddy: there, hovering over the fused glass and plastic, feeding it reams of paper and wrist-thick cartridges of ink, it occurred to me, I'm sure, that I was one step closer to revolution, usurping corporate technological resources to spread the word.

Transplanted to the punk rock mecca that is the Bay Area, I took advantage of the Punk Rock Discount—i.e., getting free stuff from peers gainfully employed in various retail shops—at a suburban Kinko's where the midnight shift worker was One of Us. She cheerfully charged eighty-four cents for what was easily a hundred dollars' worth of copying. Around midnight we'd trickle in, fresh off the Muni, clutching cut-and-paste layouts and a wide range of bottled poisons (forty-ouncers of bad beer or organic vegan smoothies), set up shop at the machines, chat, swap finished products before we'd disappear back into the gloom of strip-mall fluorescence, singly or in groups, hauling our booty in hijacked cardboard boxes. Later I made friends with a lustful punk boy who worked full-time in the copy store a block from campus; his first romancing move, in an attempt to win me over, was to let me walk out with a big stack of freshly made 'zines. His devotion translated into unlimited access to free (the magic word) copies—a Punk Rock Dream Come True, this meant that the three A.M. bus rides and long hikes up steep San Francisco hills with my loot, cradled in bone-weary arms, were a thing of the past. While such great lengths added a certain rugged, obsessive DIY (do it yourself) flair to the process, even punks sometimes like convenience.

More than simply dictating my schedule, my 'zine *Slant* ordered my life; my product was my baby and I was having trouble differentiating between it and me. Although the content of my initial forays into 'zine-age was standard anarcho-feminist stuff—smash the state, fight war not wars, abortion on demand, et cetera, nothing spectacular—I was hooked on the medium, not just the message, and so my fascination with the machine completed the queerly conceived love triangle. I wrote constantly, researched articles, hawked my wares to distributors and record stores across the country, spent late nights cutting and pasting. The skin between my fingers turned a nasty shade of gray. I was always distracted by prophetic visions of future manifestos I'd compose, magnificent covers I'd design. I often made art for 'zines I never did finish; I still have a file cabinet full of clippings I *swore* I'd use someday, fat envelopes stuffed with newspaper articles, patterned paper, and Xeroxed graphics. And I wasn't the only one obsessed with low-tech paper product; it was a communal punk malaise, a widespread phenomenon of kids hooked on glue sticks, typewriter ribbon, and paper cuts, and what they made possible. The local post office was a home-away-from-home to several other 'zinesters; we'd stop to chat about our latest issues while picking up letters, trades, and unsolicited contributions. Really, we hardly spoke of anything else. So it was my 'zine that afforded me community—a good number of friends I met (some I still haven't met, yet) years ago, when we'd trade 'zines in the mail, sparking long-term correspondence or promises of road-trip visits. It was through this incestuous network that I even met my boyfriend, also an obsessive 'zine-maker.

It was a community that felt intimate and, at the time, real, importantly facilitated by a mutual preoccupation with 'zines and Xerox machines. But soon, not long after I met the counter boy with his promise of free copies, I decided to quit punk rock. Punk rocked high on the Richter scale when Bikini Kill's Kathleen Hanna screamed, "SUCK MY LEFT ONE!" I, along with hundreds of other girls, raptly mouthed the rage-filled words at a moved and inspired sixteen. Having learned my lesson from riot grrrl's reception in a hostile and often violently defensive male-dominated scene, I'd never quite swallowed whole the foundation myth of punk egalitarianism. Still, I admit I didn't expect its lie to stick so hugely in my throat when my time to scream came. It was disappointing, but not a surprise.

There was, in the end, a small if significant revolution, girl-style even. Riot grrrl delivered a good swift kick to the masculinist punk rock paradigm where it counted most, marking the not-so-generic-after-all "whitestraightpunkboy" and confronting the popular illusion of some abstract, transcendent punk "essence." And on riot grrrl's heels, the core soon went queer too, but the race riot I wanted was clocking in at a very *very* distant third. I wonder if it's ever gonna burn *anything* down, let alone nail the punk rock paradigm to any walls. Punk rock's promise of horizontal subcultural citizenship and community was left unfulfilled.

Case in point: Some years back, the punk magazine *Maximum-RocknRoll* (*MRR*) ran a column detailing the writer's office masturbation fantasies in which Asian women figured prominently, something about how the eyelids, without the double fold, resemble vulva, inspiring erections in horny straight white men, especially this one in particular. Being a naïve nineteen-year-old anarchist feminist with a 'zine that threatened, my younger self responded with an exhaustive mini-manifesto that was subsequently published in *MRR*. I'm quite sure I invoked the Hottentot Venus, exploitative colonial relations, and "othering" discourses of exoticism and deviant sexualities.

The counterresponse consisted of a two-page column derailing me as a smarty-pants know-nothing who was probably ugly, since I obviously had no sense of humor. This dropped my desirability factor to new lows and canceled out the inherent fuckable-ness I would have otherwise earned as an Asian female. He coupled this response with a song, an Orientalized fantasy about raping me. This, the not-unexpected blow, struck a chord deep in my gut that was just a little too familiar.

So much for punk's famed egalitarianism. One person, a queer white man, quit the magazine in protest, though his dissatisfaction had been building for some time. Others, queer white women, straight Asian men, coworkers at the collectively run record store in the Mission whom I might've expected to say something, *anything* vaguely supportive, ignored the whole situation. (Here's where I think, "Community? Yeah, right.")

Years later I'm a cynic. I've shed the particular version of angry grrrl I used to play for another. I leave one foot planted in punk but my head is turned in another direction. I'm looking elsewhere to spread the word. Back in the Bay Area with no desire to renew my local p-rock membership except by proxy and, having entertained a brief illusion of hacker potential back in my teens, I return to cyberspace to check the recent hype. It's a lot different than I remember, but after a few months kicking around the Net, I tell myself, *easy*. See, somewhere along the way my Xerox love story faltered. I'm tired of cruising print shops for that new beau, another big dumb machine. The courting process bores me: the flirting involved in coaxing grayscale double-sided copies to collate properly doesn't seem worth the effort anymore. The aura of "community" that used to draw me to fanzines as do-it-yourself communication technology has faded considerably, something about how the subcultural logic of belonging commits all kinds of violence in order to police its borders. I still produce my own 'zines, if on a much more erratic schedule, and I write for a "major" punk magazine called *Punk Planet*, so it's not as if I've cut that foot from my social body. But my'zine-age takes on another dimension: *Slant* gets a phantasmatic twin, slapped together using basic HTML skills, stolen code, and pilfered graphics, existing somewhere in between my hard drive and outer space. But there it takes on new dimensions, stretching beyond now-limiting punk parameters, becoming for me an exercise in exporting theory through fiberoptic cable in the guise of a "Web journal," dated and everything. As much as I enjoy the often-painstaking process of laying out text and carefully clipped graphics on paper (and I swear I do), the convenience of the Internet, HTML templates, FTP, and the immediacy of Web publishing definitely appeal.

It's not so different, really, except for the reach in readers and the ease with which I reach them; I imagine a lot more people have read my site than my 'zines, and a smaller bunch of them emerging from a punk rock scene. No doubt many more of those who've read my site are also people of color, a phenomenon I admit I appreciate and certainly can't say for my p-rock paper audience. But content-wise, I make the same forays into cultural politics. Whereas before I drew my examples from punk rock, wielding theory in the text of a 'zine review; I depend now on what counts as "the everyday" for fodder, anything from the Spice Girls to the Fantasy Goddess of Asia Barbie™ to a Good Vibrations dildo display. And the ways I write have hardly changed; sometimes I think the language I use in cyberspace is more theoretical, and then I remember that one of the first times I'd ever read the word "hegemony" was in a grrrl 'zine. It's still a very punk rock assemblage of short essays, personal notes, and line art I've scanned and uploaded—the stuff that made up my 'zine in paper form. *Slant* still, but a little more grown-up.

exoticize this!

But wait, there's the seed of a theme, or two, here. A well-worn story repeats itself when I made the move from paper to pixels, exchanging one prosthesis (the Xerox machine) for another (my computer). I spend those same predawn hours, once occupied in front of a whirring, clicking desk-size machine, perched in front of the glowing screen of my laptop, pounding out HTML code and dancing to Sleater-Kinney, rarely returning to my stash of glue sticks. I had expanded my range of mediums and technological aids, being the type of girl who likes to have her fingers in as many pies as possible, all toward the eventual goal of multimedia empire and pop-(culture) star fame. Social upheaval remains the bonus, of course. But what actually changes in this move?

While the low-tech machinations of Xerox machines and cut-and-paste assemblage are second nature, I don't pretend to be intimately familiar with all the implications of electronic culture, what with its interactive whatever and virtual anything. What this essay will *not* do is explore all the potential implications of communications technologies as social control, demon offspring of the U.S. military-industrial complex, or expensive Silicon yuppie toy; chart hacker hierarchies and coded social practices; or map the slippery machinations of transnational corporate practice(s) and flexible capital. My skills extend only so far as my HTML scripts, which isn't all that far; I never enter chat rooms, frolic in multi-user dungeons, or post to bulletin boards. I have no particular interest in looking for "community" in the nowhere-ness of cyberspace. After a few hours updating my sites every week, I think I've already spent way too much time in front of the screen. Hardly a Luddite, I still make a pretty confused cyborg, not quite sure what to do with all these extraneous wires and circuit boards cluttering me.

Both punk rock and cyberspace prove to be vexed territories in which my civil participation takes on multiple dimensions in time, space, and theory and in political import (plotted, as I am, along certain valances: Asian/American, feminist, bi-queer). The promise of abstract citizenship offered by punk and the Internet to a decidedly not-so-abstract me is a one-handed affair, contingent, it seems, on how I want to narrate my raced, sexed, and gendered body into nominally democratic publics. Exploring what Lauren Berlant calls the "dialectic between abstraction in the . . . public sphere and the surplus corporeality of racialized and gendered subjects," I've been wanting to pick apart what it means to be visible in the body I occupy/narrate within these supposedly postnational sites and examine the consequences of my technologically enhanced Asiatic geekgrrrl agitprop.[1]

Let me set the scene, almost two years ago now. I'm dredging search engines for Asian/American feminist work and I'm coming up with, well, not a whole lot. I've run the gamut of "spiders" crawling the Web for material and after fifty screens of NAKED! ASIAN! PUSSY! sites I'm lucky if I find a link to an Asian American studies course at some community college invoking Amy Tan. I'm not surprised. The seductive selling point of the information superhighway as a vast network of deep data reservoirs, virtual libraries chock full of fun facts and lengthy edifying works, falls short of its mark, implying something, perhaps, about the demographic constituency of most Web producers. The feminist performance theorist Peggy Phelan writes, "If representational visibility equals power, then almost-naked young white women should be running Western culture."[2] It seems I might be able to offer a parallel observation, that if representational visibility equals power, then almost-naked young Asian women should be running a very big chunk of cyberspace. That is, whenever I type "asian+women" into search fields I get almost nothing but them.

Because I'm a (masochistic) project girl, I take it upon myself to respond to this particular drought. Poring through the Web to compile another Web site, this one an Asian/American feminist resource listing, I invest long hours coming up with as many possible combinations of "asian" and "women" I can, conjuring increasingly obscure variations on a theme. I find myself making use of the liberal logic of visibility politics, thinking to make "visible" the presence of Asian/American feminisms and feminist cultural production—what might otherwise be ignored or denied. This makes me nervous; I have a theoretical bone to pick with the notion that if "we" (a collectivity I invoke with a shrug) make enough of a fuss, "we" will be lucky enough to be recognized by a dominant group of "others," a power relationship that rubs me a little raw. But then again, I do have a bit of the exhibitionist in me, and I tell myself it's better, in this case, to be seen, if only for the sake of those similarly beleaguered in their quest of Asian/American feminist links on the Web.

By way of myth making/explanation, in the (second) introduction to *exoticize this!* I write,

> exoticize this! . . . is an extensive and always expanding feminist listing of resources, both on and off-line, for south asian and a.p.i. american women. originally spawned from my frustration with the web, it continues to grow for much the same reason. of course, searches for "asian + women" on various engines pull up hundreds of unimaginative hetero-porn and the ubiquitous mail-order bride sites. but more, as i pored through indices for "asian american resources" i found that women's sites—let alone ones that were self-identified as feminist—were few and far between. and, on the equally disturbing other hand, investigating the thousands of feminist sites and listings yielded little in the way of links or resources for women of color, let alone asian american women. so between asian american link sites—male-dominated and predominantly straight—and feminist ones—white-dominated and also predominantly straight—there's little intersection anywhere.

The first version of *exoticize this!* was an HTML catastrophe; my meager design skills were both meager and hardly skills. Having improved my skills slightly, I try to redesign often, although I'm not equipped for a full-scale graphic extravaganza. My content was better,

Image from *exoticize this!* Mimi Nguyen.

and is still improving, including links to and lists of Asian American feminist video- and filmmakers, artists, academics and scholars, articles, political activists and organizations. I try to pick and choose among personal Web sites for those that aren't afraid to "confess" to feminist or radical sentiments, and I admit a soft spot for my pop culture section, in which I get to flex my feminist cultural studies muscles. It's a labor of love, no doubt about it, and sometimes the love *does* feel like labor; it's true, the hours spent on updating the site start to wear a little thin on my political conviction.

And at the time I began the site I wondered what kind of potential (*assured*) mess I'd gotten into, thinking about my vague unease with being public, something I learned from punk rock. I've had to remind myself again that I like to fight, or, at least, I'm used to it.

Live Asian Girls Dissect Cyberspace Politics

The email I get runs the gamut. Teenaged Chinese American riot grrrls, South Asian lesbian activists, Minnesotan radical librarians, and white Texan graduate students make my day, regularly sending encouragement and new links my way. But there's a flip side, of course; after all, "[v]isibility is a trap . . . ; it summons surveillance and the law; it provokes voyeurism, fetishism, the colonialist/imperial appetite for possession."[3] And so I get my more-than-fair share of hate email. The feminist film theorist Laura Mulvey defined voyeurism as the will to punish the (female) threat in order to reassert control, and my detractors are hardly exempt from that particular dynamic.

Subj: Ah so, grasshopper
Date: ———
From: TabascoLuv
To: Critchicks

Hmmm, did you forget to take your prozac this morning? Now seriously, has it dawned on you that you may have severe neurotic tendencies that need to be therapeutically treated? By the way, congenital lesbians usually don't harbor a deep hatred of men! Were you molested or raped when you were younger?

It's just that I hate to see a foxy asian babe go to waste ;)

'Nuff said,
Scott[4]

This one just skims the surface; so far I've received about two dozen examples of similarly hostile responses. They make for an amusing side note on any given day, along with unsolicited personal photographs of losers cybercruising for dates and angry denunciations of my projects. I've been accused of being a man-hating unhappy lesbian (definitely unoriginal) guilty of a nefarious plot to install myself as the head of my own cult (this, now, is a new one). There are also the usual accusations of being racist, sexist, and otherwise bigoted toward white straight men. Quite a number expressed outrage over what was interpreted as unchecked violence from unexpected corners (being me). My original "splash page," a kind of front-door screen that leads to my main index, floated a pair of carved wooden chopsticks against a stark white background; the text read, "with these you could kill a man. *do you wanna?*" Apparently, some (again, most of them self-identified as white straight men) understood this as wholesale advocacy on my part that willing readers—perhaps as initiates into my cult?—randomly brain hapless men with a pair of chopsticks through the (thick) skull.

All such missives share a singular purpose, that is, to fuck with me, if only in a metaphoric sense. The lack of photographs, of me, at least, had been a deliberate omission. (Other personal sites are littered with 'em.) I have no pressing desire to cater to *that* particular urge of the anonymous voyeur, powered by a deceptively friendly neighborhood browser. (Nothing to see here, folks.) Still, my body is invoked. That is, everyone brings it up, even in its conspicuous (visual) absence, or perhaps because of it. Voyeurs are used to mastering the spectacle through its surveillance; there's a desire to create/narrate one—a maligned body, that is—even if it's imagined. I suppose it doesn't matter if it's actually mine or not; I've got an iconicity that has nothing to do with me and everything to do with a powerful social script that wants to do certain things *to* me. That is, my body can probably be envisioned/pieced together easily enough: a digitized Frankenstein amalgamation of parts seen on All Asian Action or Hot Oriental Babes. The invocation of my body serves a purpose, then, something about keeping me in line; there are the threats of sexual violence, thinly veiled suggestions that they might enjoy the scene of my particular body in pain.

Even sight/site unseen, there's something fleshy lurking in between the limits set by <html> and </html>. I mean, I enact my various identifications—including my Vietnamese-ness—deliberately, scripting these into the hypertext I write. It seems, however, that I have violated some rules regarding my in/visibility. First, in declaring my corporeal politicization, I've renounced the popular cybergeek assumption decreeing the obsolescence of (the particularities of) the body. But worse, since I've declared myself to be an Asian girl, it is unthinkable that I'm not then a specular delicacy, a visual treat to stimulate lower male regions, since most of my (*ahem*) peers—found under "asian+women"—are presented as such, bearing cleavage and abundant (and bad) lingerie. Being both embodied and abstract, still (they tell me) I've messed up the order.

While I'm no stranger to tech stuff, there's something about technologies like cyberspace that makes me wary of claiming them as my own. I don't believe I belong in the intangible *there*-ness (as in, distanced from wherever it is I am) of the Net; it doesn't make me feel at home. Cyberspace as the new national public sphere, a positive site of sociality and communication, is conditional; virtual democracy with the emphasis on *virtual*: as in *close, but no cigar*. That is, while we may all be able to operate a certain "eye of power," an authoritative vision by way of Netscape, Explorer, and various search engines, safely ensconced behind the anonymity of fiberoptic cable, we do not all have equal access to the organizing power behind the "eye."

I actually resent the offer of abstraction that cyberspace extends; it tells me there is something, after all, wrong with me. The insinuation is that it's my body—and not the cultural logic that organizes and disciplines my body—that checks my access into allegedly democratic publics, "America," punk rock, or cyberspace.

So if I narrate myself as a bisexual Vietnamese refugee-tomboy in a Bettie Page haircut (which, I suppose in a general sense, loosely maps me), am I simply unimaginative? Am I asking for it—the harassment, the freaks with their threats and insults? The suggestion that I could "pass" if I wanted to, that I disrupt the smooth flow of cyberdemocracy by "making an issue" of my specificities, has passed through my emailbox more than once. When the ideal of bodily abstraction is made readily available—again, everywhere promised but fufilled only in this technologically enhanced nonspace, where solely prosthetic bodies (email addresses, personal "homepages") circulate—it's virtually unbelievable that I might refuse.

My refusal is an assertion of my politics, of bodily histories and historicities. That is, while I wouldn't say that the body determines subjectivity, I won't pretend that the body isn't lived, interpreted, subjectively. There are, after all, the psychic and material costs of hegemonic corporeal logics—those that seek to discipline the bodies of Others, including mine—that are intimately *felt* and thus can hardly be ignored. The hostile responses to my sites, then, are perfect examples; the hierarchization of bodies in cyberspace continues apace. It's no accident that my Net-harassers make a point of not only willing my body into view, but also identifying themselves as white straight men. There's tacit awareness of the power and privilege attached to these corporeal specificities, suggesting that even *these* guys know that the premise of abstract citizenship is a joke, only one of many disciplining tricks in the bag.

Because the corporeal signs—skin, hair, features—we count on to regulate "being" and social belonging don't *physically* appear (in the organic sense) in the cybernetic void, they are, by necessity, *narrated*. Never mind the supposedly expansive hypertext-mediated encounter, the social logics with which we come to the table, or terminal, are still intact. We perform them as scripts, CGI or not, writing "legible" bodies into the electronic ether in order to clarify power relations, if for different political purposes: theirs, to reassert power, mine, to illuminate (otherwise unmarked) privilege and take power apart.

Still, in constructing these sites, I'm not consciously working with any coherent concept of "community"—summarily determined by a shaky identity politics—in mind; I mean, the instrumental use of the label "Asian American," while especially strategic in the case of *exoticize this!* feels cramped to me. I constantly worry about what I have to look like, or, in the case of cyberspace, what I have to act like, in order to fit under its supposed umbrella sign; I'm forced to invoke "community": with quotation marks always already intact. So instead I like to think that my sites are ideologically accessible to anyone similarly inclined; I imagine an audience that doesn't necessarily look like me, but *does* want to engage in a critical dialogue with me. In the end, it's about both making the assumed absent (Asian/American feminisms) present and embodying an oppositional field of electronic positions in spite, literally, of the simultaneous demand for abstraction and the subsequent policing of unruly bodies.

Conclusion

> I would like to insist on the embodied nature of all vision, and so reclaim the sensory system that has been used to signify a leap out of the marked body and into a conquering gaze from nowhere. This is the gaze that mythically inscribes all the marked bodies, that makes the unmarked category claim the power to see and not be seen, to represent while escaping representation.
>
> —Donna Haraway, *Simians, Cyborgs, and Women*

While I'm not a transcendent girl, I *am* a stubborn one. I'm not a fan of the Cartesian feint, that old mind-body split, if not a fan of an "innocent" identity politics either. There's a strategy for intervention/interaction in here somewhere, one that makes use of what's been called strategic essentialism, a necessary fiction, or even a necessary *error*, but I want to keep that in tension with a fervent belief in the impossibility of coherency, wholeness, organicity. It's how I've managed to wade through the muck of punk rock and cyberspace—

(outer) spaces that pretend to post-nationality, abstract citizenship, and other lies your Uncle told you. And if it seems tricky, well, it is. So I make a spectacle of myself knowing that in/visibility is a trap; even if I don't conjure revolution, I can nail a few paradigms to the wall, or at least make some people very uncomfortable.

Everyone cites Sandy Stone here and I'm no exception to that rule: "No refigured virtual body, no matter how beautiful, will slow the death of a cyberpunk with AIDS. Even in the age of the technosubject, life is lived through bodies."[5] It's the return of the repressed, of course. Not to suggest that the body will have its say, in the end, in the sense of the body as "natural" or pre-discursive; but that those dominant corporeal logics that *constitute social hierarchies and otherwise regulate the body will*. If, at the end of the day, I'm forced to summarize a lesson, it would be this: that cyberspace admits (with a shrug) that race, gender, sexuality, nationality are socially constructed systems of classification. It makes obvious that discourse fundamentally constitutes our social realities, *even as* the ideal model of bodily abstraction and its liberal democratic possibilities continues to exert overwhelming rhetorical charm. "Officially" endorsing this (national) ideal, American-style, cyberspace sells itself as the "final frontier" of participatory liberal citizenship where an apolitical and ahistoricizing "color blindness" (supposedly) really exists in the absence of bodies. In reality all kinds of bodies and their doubles—digitized, prosthetic, virtual, textualized—are circulated, exchanged, and performed in the electronic market because bodies do matter, at least when it comes to asserting social hierarchies and variously hegemonic cultural logics.[6] Even when gender, sexuality, race, whatever, become obvious as effects of discourse and coded information, it doesn't necessarily upset the continued ordering of differentiated bodies in public *or* the naturalization of that organization.

And here I make another confession: I don't love my laptop. Maybe I like my machines big and stupid, hulking mechanical clunkers with gears and cogs and stuff that needs to be oiled, coaxed into performing. I miss my love-object but there's no help for it, there's nothing, I suppose, like your first (for me, the Xerox machine, with punk rock attached). It's terrible, really, that I'm not all that attached to my prostheses, though I admit affection for my virtual body double. But there are only so many stories you can really tell about sitting in front of your computer, typing, before it gets to sound a bit repetitive. Somehow I can't see myself saying, "Look, I was in this really exciting chat room . . ."

But more important, I like my body, and I'd like to keep it.

NOTES

1. Lauren Berlant, "National Brands/National Body: Imitation of Life," in *The Phantom Public Sphere*, ed. Bruce Robbins (Minneapolis: University of Minnesota Press, 1993), 173–208, 178.

2. Peggy Phelan, *Unmarked: The Politics of Performance* (London: Routledge, 1993), 10.

3. Ibid., 6.

4. This is an actual message, I swear.

5. Allucquere Rosanne Stone, "Will the Real Body Please Stand Up? Boundary Stories about Virtual Cultures," in *Cyberspace: First Steps*, ed. Michael Benedikt (Cambridge: MIT Press, 1992), 112.

6. Judith Butler, *Bodies That Matter: On the Discursive Limits of "Sex"* (New York: Routledge, 1993).

Chapter Eleven

The Virtual Barrio @ the Other Frontier
(or The Chicano Interneta)

Guillermo Gómez-Peña

> [Mexicans] are simple people. They are happy with the little they got. . . . They are not ambitious and complex like us. They don't need all this technology to communicate. Sometimes I just feel like going down there & living among them.
>
> —Anonymous confession on the Web

Tecnofobia

My laptop is decorated with a 3-D decal of the Virgin of Guadalupe. It's like a traveling altar, office, and literary bank, all in one. Since I spend 70 percent of the year on the road, it is (besides the phone of course) my principal means to remain in touch with my beloved relatives and colleagues, spread throughout many cities in the United States and Mexico. Unwillingly, I have become a cyber-vato, an information superhighway bandido. Like that of most Mexican artists, my relationship with digital technology and personal computers is defined by paradoxes and contradictions: I don't quite understand them, yet I am seduced by them; I don't want to know how they work, but

This chapter is reprinted from *Clicking In: Hot Links to a Digital Culture*, edited by Lynn Hurshman-Leeson.

I love how they look and what they do; I criticize my colleagues who are acritically immersed in new technology, yet I silently envy them. I resent the fact that I am constantly told that as a "Latino" I am supposedly culturally handicapped or somehow unfit to handle high technology; yet once I have it right in front of me, I am propelled to work against it, to question it, to expose it, to subvert it, to imbue it with humor, linguas polutas—Spanglish, Frangle, gringonol, and radical politics. In doing so, I become a sort of Mexican virus, the cyber-version of the Mexican fly: tiny, irritating, inescapable, and highly contagious. Contradiction prevails.

Over a year ago, my collaborator Roberto Sifuentes and I bullied ourselves into the Net, and once we were generously adopted by various communities (Arts Wire and Latino Net, among others) we started to lose interest in maintaining ongoing conversations with phantasmagoric beings we had never met in person (that, I must say, is a Mexican cultural prejudice—if I don't know you in person, I don't really care to talk with you). Then we started sending a series of poetic/activist "techno-placas" in Spanglish. In these short communiqués we raised some tough questions regarding access, privilege, and language. Since we didn't quite know where to post them in order to get the maximum response, and the responses were sporadic, casual, and unfocused, our passion began to dim. Roberto and I spend a lot of time in front of our laptops conceptualizing performance projects that incorporate new technologies in what we believe is a responsible and original manner, yet every time we are invited to participate in a public discussion around art and technology, we tend to emphasize its shortcomings and overstate our cultural skepticism.[1] Why? I can only speak for myself. Perhaps I have some computer traumas. I've been utilizing computers since 1988; however, during the first five years, I utilized my old "lowrider" Mac as a glorified typewriter. During those years I probably deleted accidentally here and there over three hundred pages of original texts that I hadn't backed up on disks, and thus was forced to rewrite them by memory. The thick and confusing "user-friendly" manuals fell many a time from my impatient hands; and I spent many desperate nights cursing the mischievous gods of cyberspace and dialing promising "hotlines" that rarely answered.

My bittersweet relationship to technology dates back to my formative years in the highly politicized ambiance of Mexico City in the 1970s. As a young "radical artist," I was full of ideological dogmas and partial truths. One such partial truth spouted was that high technology was intrinsically dehumanizing; that it was mostly used as a means to control "us" little techno-illiterate people politically. My critique of technology overlapped with my critique of capitalism. To me, "capitalists" were rootless corporate men who utilized mass media to advertise useless electronic gadgets, and sold us unnecessary apparatuses that kept us both eternally in debt and conveniently distracted from "the truly important matters of life." These matters included sex, music, spirituality, and "revolution" California style (in the abstract). As a child of contradiction, besides being a rabid antitechnology artist, I owned a little Datsun and listened to my favorite U.S. and British rock groups on my Panasonic *importado,* often while meditating or making love as a means to "liberate myself" from capitalist socialization. My favorite clothes, books, posters, and albums had all been made by "capitalists," but for some obscure reason, that seemed perfectly logical to me. Luckily, my family never lost their magical thinking and sense of humor around technology. My parents were easily seduced by refurbished and slightly dated American and Japanese electronic goods. We bought them as *fayuca* (contraband) in the Tepito neighborhood, and they occupied an important place in the decoration of our "modern" middle-class home. Our huge color TV set, for example, was decorated so as to perform the double function of entertainment unit and involuntary postmodern altar—with nostalgic photos, plastic flowers, and assorted figurines all around it—as was the sound system next to it. Though I was sure that with the scary arrival of the first microwave oven to our traditional kitchen our delicious daily meals were going to turn overnight into sleazy fast food, my mother soon realized that *el microondas* was only good to reheat cold coffee and soups. When I moved to California, I bought an electric ionizer for my grandma. She put it in the middle of her bedroom altar and kept it there—unplugged of course—for months. When I next saw her, she told me, "Mijito, since you gave me that thing, I truly can breathe much better." And probably she did. Things like televisions, shortwave radios, and microwave ovens, and later on ionizers, Walkmans, calculators, and video cameras were seen by my family and friends as high technology, and their function was as much pragmatic as it was social, ritual, and aesthetic. It is no coincidence then that in my early performance work, technology performed both ritual and aesthetic functions.

Verbigratia

For years, I used video monitors as centerpieces for my "techno-altars" on stage. I combined ritualistic structures, spoken word multilingual poetry, and activist politics with my fascination for "low-tech." Fog machines, strobe lights, and gobos, megaphones and cheesy voice filters have remained since then trademark elements in my "low-tech/high-tech" performances. By the early 1990s, I sarcastically baptized my aesthetic practice "Aztec high-tech art," and when I teamed with Cyber-Vato Sifuentes, we decided that what we were doing was "techno-razcuache art." In a glossary that dates back to 1993, we defined it as "a new aesthetic that fuses performance art, epic rap poetry, interactive television, experimental radio and computer art; but with a Chicanocentric perspective and a sleazoid bent."

> (El Naftaztec turns the knobs of his "Chicano virtual reality machine" and then proceeds to feed chili peppers into it. The set looks like a Mexican sci-fi movie from the 1950s.) El Naftaztec (speaking with a computerized voice): *So now, let's talk about the TECHNOPAL 2000, a technology originally invented by the Mayans with the help of aliens from Harvard. Its CPU is powered by Habanero chili peppers, combined with this or DAT technology, with a measured clock speed of 200,000 megahertz! It uses neural nets supplemented by actual chicken-brain matter and nacho cheese spread to supply the massive processing speed necessary for the machine to operate. And it's all integrated into one sombrero! Originally, the Chicano VR had to use a poncho, but with the VR sombrero, the weight is greatly reduced and its efficiency is magnified. And now, we have the first alpha version of the VR bandanna dos mil, which Cyber-Vato will demonstrate for us!* (Cyber-Vato wears a bandanna over his eyes. It is connected by a thick rope to a robotic glove. Special effects on the TV screen simulate the graphics and sounds of a VR helmet.)
>
> —From "Naftaztec," an interactive TV project about Mexicans and high technology

The mythology goes like this. Mexicans (and other Latinos) can't handle high technology. Caught between a preindustrial past and an imposed postmodernity, we continue to be manual beings—*homo fabers* par excellence, imaginative artisans (not technicians)—and our understanding of the world is strictly political, poetical, or metaphysical at best, but certainly not scientific. Furthermore, we are perceived as sentimental and passionate, meaning irrational; and when we decide to step out of our realm and utilize high technology in our art (most of the time we are not even interested), we are meant to naively repeat what others have already done. We often feed this mythology by overstating our romantic nature and humanistic stances and/or by assuming the role of colonial victims of technology. We are ready to point out the fact that "computers are the source of the Anglos' social handicaps and sexual psychosis" and that communication in America, the land of the future, "is totally mediated by faxes, phones, computers, and other technologies we are not even aware of." We, "on the contrary," socialize profusely, negotiate information ritually and sensually, and remain in touch with our primeval selves. This simplistic binary worldview presents Mexico as technologically underdeveloped yet culturally and spiritually overdeveloped and the United States as exactly the opposite. Reality is much more complicated: the average Anglo-American does not understand new technologies either; people of color and women in the United States clearly don't have equal access to cyberspace; and at the same time, the average urban Mexican is already afflicted in varying degrees by the same "first world" existential diseases produced by advanced capitalism and high technology. In fact the new generations of Mexicans, including my hip generation-Mex nephews and my seven-year-old fully bicultural son, are completely immersed in and defined by personal computers, video games, and virtual reality. Far from being the romantic preindustrial paradise of the American imagination, the Mexico of the 1990s is already a virtual nation whose cohesiveness and boundaries are provided solely by television, transnational pop culture, and the free market. It is true that there are entire parts of the country that still lack basic infrastructures and public services (not to mention communications technology). But in 1996 the same can be said of the United States, a "first world" nation whose ruined "ethnic" neighborhoods, Native American reserves, and rural areas exist in conditions comparable to those of a "third world" country. When trying to link, say, Los Angeles and Mexico City via video-telephone, we encounter new problems. In Mexico, the only artists with "access" to this technology are upper-class, politically conservative, and uninteresting. And the funding sources down there willing to fund the project are clearly interested in controlling who is part of the experiment. In other words, we don't really need Octavio Paz conversing with Richard Rodriguez. We need Ruben Martinez talking to Monsivais, as well.

> The world is waiting for you—so come on!
> —ad for America Online

The Cyber-Migra

Roberto and I arrived late to the debate. When we began to dialogue with artists working with new technologies, we were perplexed by the fact that when referring to cyberspace or the Net, they spoke of a politically neutral/raceless/genderless/classless "territory" that provided us all with "equal access" and unlimited possibilities of participation, interaction, and belonging—especially belonging. Their enthusiastic rhetoric reminded us of both a sanitized version of the pioneer and cowboy mentalities of the Old West ("Guillermo, you can be the first Mexican ever to do this and that in the Net"), and the early-century Futurist cult to the speed and beauty of epic technology (airplanes, trains, factories, etc.). Given the existing "compassion fatigue" regarding political art dealing with issues of race and gender, it was hard not to see this feel-good utopian view of new technologies as an attractive exit from the acute social and racial crisis afflicting the United States. We were also perplexed by the "benign (not naive) ethnocentrism" permeating the debates around art and digital technology. The unquestioned lingua franca was of course English, the "official language of international communications"; the vocabulary utilized in these discussions was hyperspecialized and depoliticized; and if Chicanos and Mexicans didn't participate enough in the Net, it was solely because of lack of information or interest (not money or access), or again because we were "culturally unfit." The unspoken assumption was that our true interests were grassroots (by grassroots I mean the streets), representational, or oral (as if these concerns couldn't exist in virtual space). In other words, we were to remain dancing salsa, painting murals, writing flamboyant love poetry, and plotting revolutions in rowdy cafes. We were also perplexed by the recurring labels of "originality" and "innovation" attached to virtual art. And it was not the nature, contents, and structural complexity of the parallel realities created by digital technology, but the use of the technology per se that seemed to be "original" and "innovative." That, of course, has since engendered many conflicting responses. Native American shamans and medicine men rightfully see their centuries-old "visions" as a form of virtual reality. And Latin American writers equate their literary experimentation with involuntary hypertexts and vernacular postmodern aesthetics, and so do Chicanos and Chicanas. Like the pre-multicultural art world of the early 1980s, the new high-tech art world assumed an unquestionable "center" and drew a dramatic digital border. On the other side of that border lived all the techno-illiterate artists, along with most women, Chicanos, African Americans, and Native Americans. The role for us, then, was to assume, once again, the unpleasant but necessary role of cultural invaders, techno-pirates, and coyotes (smugglers). And then, just as multiculturalism was declared dead as soon as we began to share the paycheck, now as we venture into the virtual barrio for the first time, some asshole at M.I.T. declares it dead. Why? It is no longer an exclusive space. It emulates too much real life and social demographics. Luckily many things have changed. Since we don't wish to reproduce the unpleasant mistakes of the multicultural days, our strategies are now quite different: we are no longer trying to persuade anyone that we are worthy of inclusion. Nor are we fighting for the same funding (since funding no longer exists). What we want is to "politicize" the debate; to "brownify" virtual space; to "spanglishize the Net"; to "infect" the lingua franca; to exchange a different sort of information—mythical, poetical, political, performative, imagistic; and on top of that to find grassroots applications to new technologies and hopefully to do all this with humor and intelligence. The ultimate goals are perhaps to help the Latino youth exchange their guns for computers and video cameras, and to link the community centers through the Net. CD-ROMs can perform the role of community memory banks, while the larger virtual community gets used to a new presence, a new sensibility, a new language.

NOTE

1. See www.sfgate.com/foundry/pochanostra.html.

Contributors

Vivek Bald is a New York–based filmmaker. His first documentary, *Taxi-vala/Auto-biography,* premiered at the Whitney Museum of American Art in 1994, was broadcast on WNET-TV's *Reel New York* in 1996, and has screened at festivals and conferences throughout the United States and in Canada, India, and Japan. *Mutiny: Asians Storm British Music,* his second full-length documentary, is about South Asian youth, music, and politics in Britain. He is also a musician and DJ (under the name Siraiki) and co-organizes a weekly club night in New York, also called Mutiny, dedicated to new music from the South Asian diaspora and beyond. More information about *Taxi-vala, Mutiny,* and the artists mentioned in his interview is available at www.mutinysounds.com.

Ben Chappell is pursuing a doctorate in anthropology from the Center for Intercultural Studies in Folklore and Ethnomusicology at the University of Texas at Austin. His M. A. thesis, "Making Identity with Cars and Music: Lowriders and Hip hop as Urban Chicano Performance," was nominated by University of Texas at Austin for the Conference of Southern Graduate Schools 2000 Master's Thesis Award. Other recent work includes "Is a Lowrider Postmodern? Critique, Politics and Ambivalence in Hybrid Performance," in *Text, Practice, Performance,* and "Folklore Semiotic: Charles Peirce and the Experience of Signs," forthcoming in *Folklore Forum.* He is currently conducting fieldwork for his dissertation, an ethnography of Texas lowriding.

Beth Coleman (M. Singe) is an artist mixing in text and sound. She is the codirector of SoundLab Cultural Alchemy and one of its resident DJs. She is completing a doctorate on digital culture, aesthetics, and identity at New York University.

Guillermo Gómez-Peña is an interdisciplinary artist and writer. He was born and raised in Mexico City and came to the United States in 1978. Gómez-Peña explores cross-cultural issues and North-South relations through performance, bilingual poetry, journalism, video, radio, and installation art. He has contributed to the national radio programs *Crossroads* and *Latino USA*, and is a contributing editor to *High Performance* and the *Drama Review*. Gómez-Peña is a 1991 recipient of the MacArthur Fellowship. He is the author of *Warrior for Gringostroika* and *The New World Border*.

A PC software veteran with over ten years of experience, *McLean Mashingaidze Greaves* combines extensive experience in print and television journalism with a long track record in the high-tech industry dating back to 1986. His written work is published internationally in *Essence, Vibe,* the *Source,* the *Globe and Mail* (Canada), and the *Quarterly Black Review of Books*. In addition, Greaves has served as an editor at *Paper Magazine, BET Weekend Magazine,* and *City-TV News* (Toronto) and at the Canadian Broadcasting Corporation as a TV correspondent and new media executive. He is the founder of the award-winning urban pioneer Web site cafelosnegroes.com and its parent company, Virtual Melanin, Inc.

Logan Hill is a staff writer at *New York Magazine* and a regular contributor to *Wired.com*. His writing has also appeared in the *Nation, NYArts, Playboy, Publishers Weekly,* and other publications.

Alicia Headlam Hines holds a master's degree in American studies from New York University. She teaches literature and language arts at the Horace Mann School in Riverdale, New York.

Karen J. Hossfeld is an associate professor of sociology at San Francisco State University. She is the author of *Small, Foreign, and Female: Portraits of Gender, Race, and Class in Silicon Valley* (forthcoming), and a contributor to *Feminist Approaches to Theory and Methodology: An Interdisciplinary Reader* and *Women of Color in American Society*.

Amitava Kumar teaches English and cultural studies at the University of Florida. He is the author of *Passport Photos* and editor of two anthologies, *Class Issues* and *Poetics/Politics*. He has written for a variety of publications, including *Race & Class, Critical Inquiry, College Literature, Rethinking Marxism,* and the *Nation*.

Casey Man Kong Lum teaches communication and media studies at William Paterson University. His research and teaching interests include media ecology, mass and telecommunications, ethnography of communication and media uses, intercultural communication, and Asian American media and culture. Lum is the author of *In Search of a Voice*, has guest-edited a special issue on "The Intellectual Roots of Media Ecology" for the *New Jersey Journal of Communication*, and is working on an anthology on media ecology.

Alondra Nelson is a Henry Mitchell MacCracken Fellow at New York University, where she is a Ph.D. candidate in the American studies program. Nelson lectures and writes on science and technology, digital art, and new media. She is the founder of AfroFuturism, an online discussion of technoculture in the African diaspora.

Mimi Nguyen is a doctoral candidate in the comparative ethnic studies program at the University of California, Berkeley, where her research focuses on transnational cultural studies and poststructuralist feminist and queer theorizing. Her dissertation work reconsiders the politics of memory, "home," and nation making in the Vietnamese diaspora, but she is otherwise preoccupied with performance, performativity, and prosthetic sociality. She is also a columnist for *Punk Planet* and spends too much time at her Web site: www.worsethanqueer.com.

Tricia Rose is an associate professor of Africana studies and history at New York University. She is the author of *Black Noise: Rap Music and Black Culture in Contemporary America* and coeditor, with Andrew Ross, of *Microphone Fiends: Youth Music and Youth Culture*. Her essays on cultural politics, black popular music and culture, and popular cultural theory have appeared in numerous journals and magazines, including *Camera Obscura, Social Text, Women's Review of Books, Critical Matrix, Journal of Popular Music and Society, South Atlantic Quarterly, Black Renaissance, Vibe Magazine, Artforum, Bookforum*, and the *Village Voice*. She is continuing work on a book that explores the history and cultural politics of black female intimacy and sexuality in America.

Andrew Ross is a professor and director of the graduate program in American studies at New York University. His books include *The Celebration Chronicles, Real Love, The Chicago Gangster Theory of Life,*

Strange Weather, and *No Respect*. He has also edited several volumes, including *No Sweat: Fashion, Free Trade, and the Rights of Garment Workers*.

Thuy Linh N. Tu is a Ph.D. candidate in the American studies Program at New York University. She lectures and writes about contemporary technoculture and Asian American popular culture. She is a contributing editor at *Indieplanet.com* and has been an editor at *A.Magazine*.

Ben Williams has lived and worked in New York City since moving from London to study at New York University. He has written on music, cyberculture, theater, film, and architecture for the *Village Voice*, *Artforum*, and *Citysearch New York*, among others.

Index